Value-Driven
PROJECT
MANAGEMENT

Harold Kerzner, Ph.D.

Frank P. Saladis, PMP

WILEY

John Wiley & Sons, Inc.

INTERNATIONAL
Institute for Learning, Inc.

For general information about our other products and services, please contact our Customer Care Department within the United States at (800) 762-2974, outside the United States at (317) 572-3993 or fax (317) 572-4002.

Wiley also publishes its books in a variety of electronic formats. Some content that appears in print may not be available in electronic books. For more information about Wiley products, visit our web site at www.wiley.com.

"PMI", the PMI logo, "OPM3", "PMP", "PMBOK" are registered marks of Project Management Institute, Inc. For a comprehensive list of PMI marks, contact the PMI Legal Department.

Library of Congress Cataloging-in-Publication Data:
Kerzner, Harold.
    Value-driven project management/Harold Kerzner, Frank P. Saladis.
       p.   cm.—(The IIL/Wiley series in project management)
    Includes index.
    ISBN 978-0-470-50080-4 (cloth)
       1. Project management.   2. Value analysis (Cost control)   I. Saladis, Frank P.   II. Title.
    HD69.P75K496   2009
    658.4'04—dc22
                                                                    2009018445

Printed in the United States of America

10 9 8 7 6 5 4 3 2 1

CONTENTS

Preface   vii
Acknowledgments   xi
International Institute for Learning, Inc. (IIL)   xii

Chapter 1:
HOW PROJECT MANAGEMENT HAS CHANGED   1
Why Traditional Project Management May Not Work   2
Today's View of Project Management   8
Changing Views of Project Management   16
Recognizing the Need for Change   46

Chapter 2:
CHANGING OUR DEFINITION OF
PROJECT SUCCESS   49
Changing Times   50
Not Meeting the Triple Constraint   52
Defining Project and Program Success   54
Redefining the Triple Constraint Success Criteria   56
Definition of Success   58

Chapter 3:
THE IMPORTANCE OF VALUE   61
Success   62
Types of Value   64
Return on Investment (ROI)   66
Types of Business Values   68
Changing Values   70

Chapter 4:
THE STAKEHOLDERS' VIEW OF VALUE 103

Stakeholder Perception   104

Classification of Stakeholders   106

The Sydney, Australia, Opera House   108

Apple's Lisa Computer   112

Denver International Airport   116

Balancing Stakeholders' Needs   120

Traditional Conflicts over Values   122

Project Management Value Conflicts   124

Value Perceptions within a Project   126

Chapter 5:
THE COMPONENTS OF SUCCESS 129

Four Cornerstones of Success   130

Categories of Success   132

Categories of Values   134

Deciding on the Quadrant   138

Internal Values   140

Financial Values   142

Future Values   144

Customer-Related Values   146

Reasons for Internal Value Failure   148

Reasons for Financial Value Failure   150

Reasons for Future Value Failure   152

Reasons for Customer-Related Value Failure   154

Antares Solutions   156

General Electric (Plastics Group)   158

Asea Brown Boveri (ABB)   160

Westfield Group   162

Computer Associates Technology Services   164

Convergent Computing   166

Motorola   168

Automotive Suppliers Sector   170

Banking Sector   172

Commodity Products (Manufacturing) Sector   174

Large Companies   176

Small Companies   178

Chapter 6:
SUCCESS AND BEST PRACTICES   181

From Values to Best Practices   182

Two Components of Success   184

Redefining Value Metrics (CSFs and KPIs)   186

The Need for Changing Metrics   188

Project Management Office Involvement   190

Discovery of Best Practices   192

The Debriefing Pyramid   194

Disclosure of Best Practices   196

Levels of Success in Obtaining Values   198

Project Management Knowledge   200

Project Management Benchmarking   202

Sharing Values during Benchmarking   204

Intellectual Property Cost versus Value   206

Implementation Failures   208

Chapter 7:
THE VALUE CONTINUUM   211

The Timing of Values   212

The Value Continuum   214

Barriers along the Continuum   216

Activities to Speed Up the Value Continuum   218

The Value Continuum and the Project
Management Maturity Model   220

Value Management Life-Cycle Phases   222

Value Identification Phase: Business Case   224

Business Drivers Phase: Business Drivers   226

Measurement Phase: Key Performance Indicators   228

Value Realization Phase: Value (Benefits)   230

Customer Satisfaction Management Phase:
Continuous Improvement   232

Chapter 8:
ASSIGNING VALUE THROUGH OBJECTIVES 235

Types of Performance Reports   236

Benefits and Value at Completion   238

Determining Benefits (Value) at Completion   240

Establishing the Business Objectives   242

Estimating Approaches   246

Project Plans   248

Business Plans   250

Canceling Projects   252

Marrying Project and Program Management   254

Chapter 9:
VALUE LEADERSHIP AND SENIOR MANAGEMENT 257

The Evolution of Leadership   258

Measurements and Triggers   260

What Executives Want to Hear   262

Critical Issues for the Selling Process   264

Threats that Executives Face   266

Project Management Success versus Maturity   268

Conclusions   270

Index   273

PREFACE

For more than 40 years, the traditional view of project management was based on a belief that if you completed the project by adhering to the well-known triple constraint of time, cost, and performance, the project was considered to be successful. Perhaps in the eyes of the project manager and possibly the sponsor, the project appeared to be a success. But in the eyes of the customer or even the parent or sponsoring company's senior management, the project might be regarded as a failure.

The changing economic climate and the increasingly competitive global environment are driving project managers to become more business oriented. Projects are now being viewed from a strategic perspective and as part of a business or enterprise for the purpose of providing value to both the ultimate customer and the parent corporation. Project managers are expected to understand business operations much more so today than in the past. Some companies have begun developing and delivering internal training programs for their project managers specifically focused on business processes. As project managers become more business oriented, the definition of project success now includes a **business component**. The business component is directly related to value.

Projects must provide some appreciable degree of value when completed in addition to meeting the objectives associated with the triple constraint. Perhaps many project managers believe that achieving the parameters of the triple constraint means providing value, but that's not always the case. Why should a company select and assign resources to work on projects that provide no measurable and documentable near-term or long-term value? Too many companies are either working on the wrong projects or simply have an inadequate project selection process. Project portfolios are filled with projects that do not provide real value at completion even though the triple constraints have been managed carefully and met.

Assigning valuable resources to projects that provide no appreciable value internally to the organization or externally to a client is an example of truly inept management and poor decision making. Yet selecting projects that will guarantee value or an acceptable return on investment (ROI) is very challenging because some of today's projects do not provide the targeted value until perhaps years into the future. This is particularly true for research and development (R&D) and new product development, where as many as 50 or more ideas must be explored to generate one commercially successful product. In the pharmaceutical industry, the cost of developing a new drug could run about $850 million, take 3,000 days to go from exploration to commercialization, and provide no meaningful return on investment. In the pharmaceutical industry, less than 3 percent of the R&D projects are ever viewed as a commercial success and generate more that $400 million per year in revenue.

There are, of course, multiple views of the definition of value. For the most part, value is viewed very similarly to how we view beauty—it is in the eyes of the beholder. In other words, value may be viewed as a perception at project selection and initiation based on data available at the time. But at project completion, the actual value becomes a reality that may not meet the expectations that had initially been perceived.

Another problem is that the achieved value of a project may not satisfy all of the key stakeholders since each stakeholder may have a different perception of value as it relates to their particular business function. The definition of value can be industry specific, company specific, or even dependent on the size, nature, culture, and business base of the firm. Some stakeholders may view value as job security or profitability. Others might view value as image, brand recognition, reputation, or the creation of intellectual property. Satisfying all stakeholders is a formidable task that is often difficult to achieve and, in some cases, may simply be impossible.

When the true value of a project is obtained, the company must decide how to capitalize on what has been gained. The projects and associated procedures that resulted in the value can either lead to or become examples of best practices that are formally documented and advertised in organizational literature. Other forms of value may be seen as company proprietary information and intellectual property that differentiates the company from competitors, the details of which are not released publicly. In any event, the ultimate goal is to define and achieve value.

HAROLD KERZNER, PH.D.
FRANK P. SALADIS, PMP
INTERNATIONAL INSTITUTE FOR LEARNING, INC., 2009

ACKNOWLEDGMENTS

Some of the material in this book has been either extracted or adapted from Harold Kerzner's *Project Management: A Systems Approach to Planning, Scheduling, and Controlling,* 10th edition; *Advanced Project Management: Best Practices on Implementation,* 2nd edition; *Strategic Planning for Project Management Using a Project Management Maturity Model*; *Project Management Best Practices: Achieving Global Excellence,* 1st edition (all published by John Wiley & Sons, Inc.).

Reproduced by permission of Harold Kerzner and John Wiley & Sons, Inc.

We would like to sincerely thank the dedicated people assigned to this project, especially the International Institute for Learning, Inc. (IIL) staff and John Wiley & Sons, Inc. staff for their patience, professionalism, and guidance during the development of this book.

We would also like to thank E. LaVerne Johnson, Founder, President & CEO, IIL, for her vision and continued support of the project management profession, Judith W. Umlas, Senior Vice President, Learning Innovations, IIL, and John Kenneth White, MA, PMP, Senior Consultant, IIL for their diligence and valuable insight.

In addition, we would like to acknowledge the many project managers whose ideas, thoughts, and observations inspired us to initiate this project.

—HAROLD KERZNER, PH.D., AND FRANK SALADIS, PMP

INTERNATIONAL INSTITUTE FOR LEARNING, INC. (IIL)

International Institute for Learning, Inc. (IIL) specializes in professional training and comprehensive consulting services that improve the effectiveness and productivity of individuals and organizations.

As a recognized global leader, IIL offers comprehensive learning solutions in hard and soft skills for individuals, as well as training in enterprise-wide Project, Program, and Portfolio Management; PRINCE2®*; Lean Six Sigma; Microsoft® Office Project and Project Server**, and Business Analysis.

After you have completed *Value-Driven Project Management*, IIL invites you to explore our supplementary course offerings. Through an interactive, instructor-led environment, these virtual courses will provide you with even more tools and skills for delivering the value that your customers and stakeholders have come to expect.

For more information, visit www.iil.com or call 1-212-758-0177.

---

*PRINCE2® is a registered trademark of the Office of Government Commerce in the United Kingdom and other countries.
**Microsoft Office Project and Microsoft Office Project Server are registered trademarks of the Microsoft Corporation.

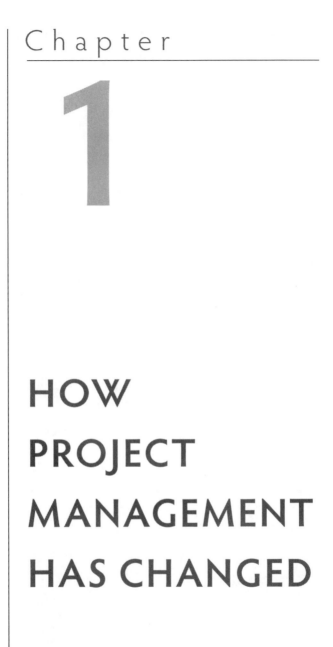

Chapter

1

HOW PROJECT MANAGEMENT HAS CHANGED

WHY TRADITIONAL PROJECT MANAGEMENT MAY NOT WORK

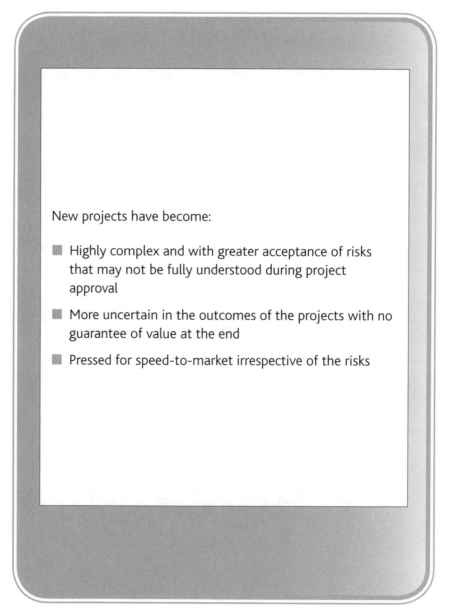

New projects have become:

- Highly complex and with greater acceptance of risks that may not be fully understood during project approval

- More uncertain in the outcomes of the projects with no guarantee of value at the end

- Pressed for speed-to-market irrespective of the risks

Traditional project management works well when the direction of the project is clearly understood, the scope is well defined, all key stakeholders agree on the objectives and expectations, the risks have been assessed and well understood, and the probability of success is considered to be very high. In comparison, for companies that wish to be innovative and become market leaders rather than market followers, the type of projects approved may be based on "fuzzy" objectives, optimism, and a willingness to take risks and basically do not follow a specific set of selection criteria.

More and more projects are highly complex and may require a technical breakthrough to achieve success. In addition, the risks associated with achieving the breakthrough can be significant, there is no guarantee that the project will be successful, and that the expected value at the completion of the project will be achieved. If a market leadership position is desired, project planning and execution are further complicated by competition and the requirement to compress the schedule for an early introduction into the marketplace.

Today's projects are not necessarily as well defined and understood as projects in the past. The global economy, rising costs, and competition are driving many companies to take greater risks to achieve their business objectives. As a result, the traditional theories of project management may not work well when applied to these new types of projects. We may need to change the way we manage and make decisions about projects. Business decisions and requirements may very well override technical decisions and project requirements.

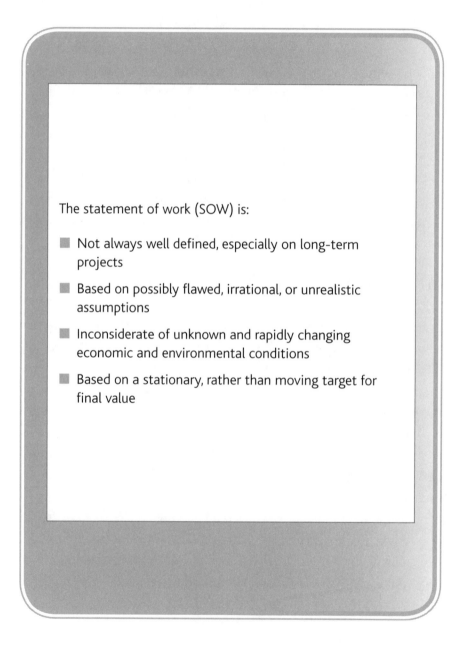

The statement of work (SOW) is:

- Not always well defined, especially on long-term projects

- Based on possibly flawed, irrational, or unrealistic assumptions

- Inconsiderate of unknown and rapidly changing economic and environmental conditions

- Based on a stationary, rather than moving target for final value

As projects become more complex, the statements of work (SOWs), in many cases, become less well defined and possibly ill defined. Typically, with all SOWs, assumptions are developed. When dealing with long-term projects, assumptions about environmental conditions and the economy are subject to considerable change and almost impossible to truly define with any sense of confidence. In such cases, the value achieved from the deliverable can be expected to become more important. Also, the achieved value may not have been fully understood initially and may have changed over the life of the project. Therefore, the final value of the project may be a moving target rather than a stationary target, and the intended customer and associated stakeholders may have to accept a deliverable with a final value that is quite different from initial expectations. The greater the project duration, the greater the chance that the final result will be significantly different from the initial objectives.

Given our premise that project managers are now more actively involved in the business, we must track the assumptions the same way that we track budgets and schedules. If the assumptions are incorrect or no longer valid, then we may be required to change the plan, change the SOW, or consider canceling the project. We should also track the project's expected value as decisions are made because these decisions may result in unacceptable changes to the final value of the project and could create a reason for project cancellation. These concerns will be discussed in later portions of the book.

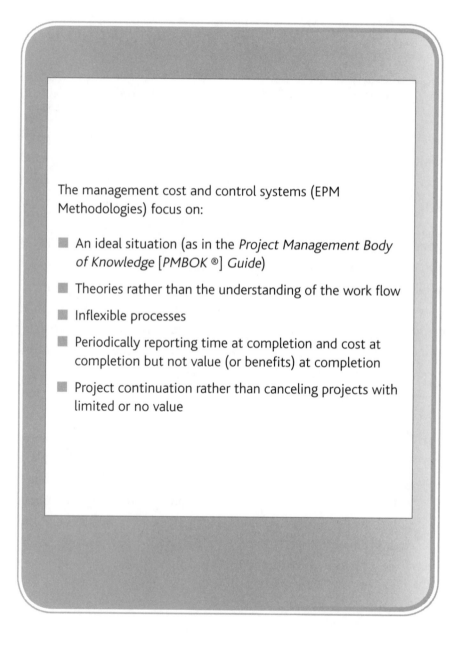

The management cost and control systems (EPM Methodologies) focus on:

- An ideal situation (as in the *Project Management Body of Knowledge [PMBOK ®] Guide*)

- Theories rather than the understanding of the work flow

- Inflexible processes

- Periodically reporting time at completion and cost at completion but not value (or benefits) at completion

- Project continuation rather than canceling projects with limited or no value

Most companies either have or are in the process of developing an enterprise project management (EPM) methodology. EPM systems are usually rigid processes designed around policies and procedures, and work efficiently when the statement of work is well defined. But with the new type of projects expected over the next decade, these rigid and inflexible processes may be more of a hindrance.

EPM systems must become more flexible in order to satisfy business needs. The criteria for good systems will lean toward forms, guidelines, templates, and checklists rather than policies and procedures. Project managers will be given more flexibility in order to make decisions necessary to satisfy the business needs of the project.

In the future, the assumption that the original plan is correct will become an increasingly poor assumption. As the providing or receiving organization's business needs change, the need to change the project plan will become evident. Also, decision making based entirely on the triple constraint, with little regard for the final value of the project, may result in extreme stakeholder dissatisfaction or significant opportunity cost.

Simply stated, today's view of project management is quite different than the views of the past, and this is partially the result of recognizing the many benefits realized through project management over the past two decades. The following illustrations show the changing views.

TODAY'S VIEW OF PROJECT MANAGEMENT

Today, we are managing our business by projects.

After more than 40 years of analysis, lessons learned, and the distribution of volumes of best practice documents; companies have come to the realization that a defined and efficiently implemented project management methodology does work and is very beneficial to an organization's growth and stability. The fundamental principles of project management can be applied to all parts of a business. Simply stated, companies are managing their business through projects, and every major activity within a company can be viewed as a project.

Project management, as a discipline, affects all parts of a business and is present at all levels of management. Each functional unit generally manages projects that support higher-level business objectives, and line managers or functional managers are being trained in project management techniques to manage projects that are exclusively within their functional area. Although project management has been viewed as a profession, it is only within the last decade or so that companies have been creating specifically designed career paths and positions for project managers.

Of significant importance is the focus on training executives to function as project sponsors. In this role, the project sponsor provides the project manager with funding and critical project information (business-related information affecting the project), which eliminates roadblocks facing the project manager. The sponsor also acts as the referee or facilitator in resolving major conflicts and problems between the project manager and other business units or functional entities.

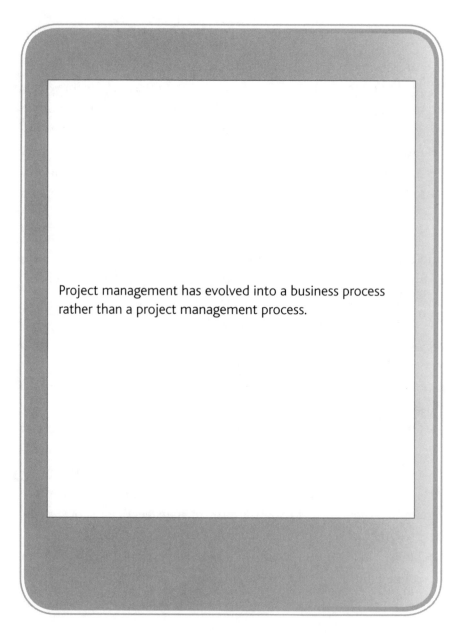

Project management has evolved into a business process rather than a project management process.

B ecause project management affects all parts of a business, it is now viewed as a business process rather than merely a methodology to meet a specific objective. This aligns itself with the current trends toward EPM.

Historically, during competitive bidding activities, contractors would emphasize only their project management processes and how they would be used to produce the deliverables to meet the customer's needs. Today, companies are integrating project management processes with their business management processes to become more efficient, promote interest, and attract new clients. Some companies have even been fortunate enough to receive single-source contracts because of the faith that the customer has in the contractor's ability to repeatedly meet deliverables.

It is important to understand that meeting the customers' requirements is sometimes accomplished through the expense of disrupting the corporate culture and ongoing business operations. Today's project manager must be knowledgeable about both the business processes and the project management processes to make the most appropriate and effective decisions in the best interest of the company and the project.

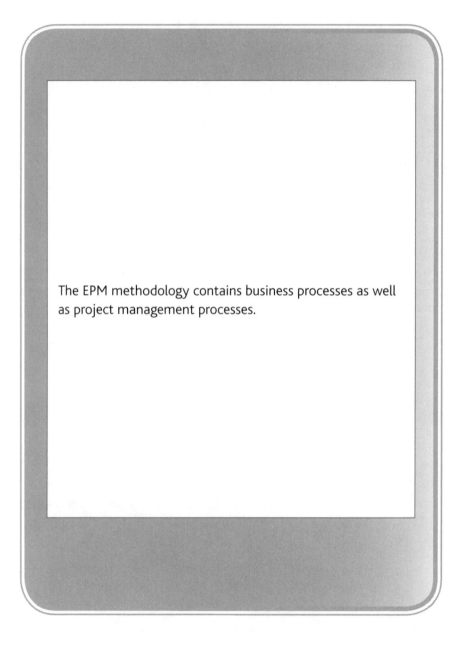

The EPM methodology contains business processes as well as project management processes.

We discussed the importance of project management being integrated into business processes. The reverse is also true in that business processes are being integrated into formal project management methodologies. Historically, project management methodologies contained the following:

- Step-by-step processes for planning and managing projects

- Concurrent engineering (improved speed to market)

- Total Quality Management (TQM) and Six Sigma

- Risk management

- Change and configuration management

Many companies have adopted an EPM methodology, which is basically one methodology selected for use by all business units to manage all projects. With an EPM methodology in place and practiced, it is fairly easy to integrate many business processes into the methodology such as:

- Supply chain management

- Feasibility studies

- Cost-benefit analyses

- Capital budgeting

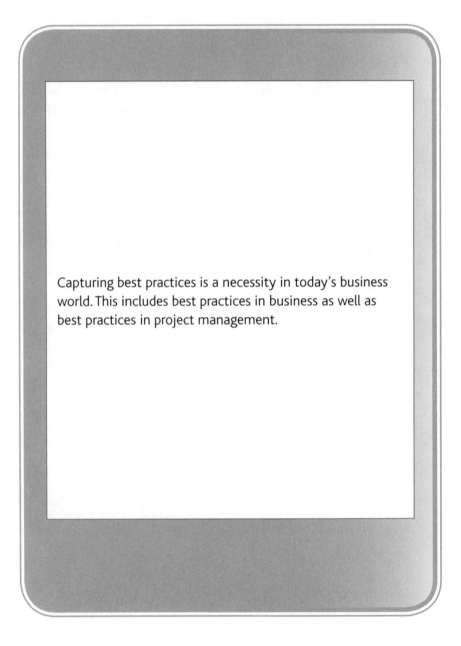

Capturing best practices is a necessity in today's business world. This includes best practices in business as well as best practices in project management.

Capturing best practices has become a business necessity. Best practices libraries are viewed as competitive weapons and can create significant advantages during the bidding process. Consider a company that issues a request for proposal (RFP) and receives identical lowest bids from two bidders. The first company has a well-maintained and utilized best practices library and is willing to share the library with the client upon contract award. The second low bidder does not maintain a best practices library. With all other things being equal, the first company would generally be awarded the contract.

Historically, project managers have been expected to capture best practices, but related to project management only. Today, because project managers are now being viewed as business managers as well as project managers, they must also capture business-related best practices. Not all of the best practices are shared with the customers, however. The business-related best practices may be viewed as proprietary knowledge and not shared externally with customers or contractors. Some companies even maintain two best practices libraries—one for the customer's benefit and one for internal use only. But, in any event, the capturing of best practices is a business necessity.

CHANGING VIEWS OF PROJECT MANAGEMENT

	Historical View	1990	Today
Project manager's role and responsibility	Monitor and control during execution	Planning for project execution	Strategy development and project selection input
When brought on board	After contract award or at end of initiation	During proposal preparation	During concept development and input in the bid/no-bid decision
Knowledge requirements	Technical knowledge (command of technology)	Mostly technical but some business knowledge	Mostly business but some technical knowledge (understanding of technology)
Customer expectations	Deliverables	Deliverables	Business solutions
Definition of success	Meeting the triple constraint	Meeting the triple constraint	Multiple success criteria (both project and business success)

Historically, project managers were viewed as organizational resources that were task oriented and did not necessarily possess a strong knowledge of business and strategic issues, but were exceptional at developing and executing plans. It was a common business practice to engage a planning or estimating group to develop project plans and then assign a project manager to execute the plan. Typically, project managers were not brought on board during project initiation and were assigned to the project after the major part of planning had been completed. Occasionally, a project manager who demonstrated significant skills in project planning would be assigned this function exclusively. After the project was planned, a second project manager would be assigned to the project to manage and plan execution. In some cases, a third project manager would be assigned to manage project closure.

Gradually, and through experience, project managers improved and fine-tuned their planning skills. Today, as project managers become more actively involved in business decisions, project manager input is requested for:

- Establishing project objectives at the technical and business component level

- Strategy development that connects project objectives with business objectives

- Project selection as part of portfolio management

- Project prioritization as part of portfolio management

- The bid or no-bid decision

	Historical View	1990	Today
Project manager's role and responsibility	Monitor and control during execution	Planning for project execution	Strategy development and project selection input
When brought on board	After contract award or at end of initiation	During proposal preparation	During concept development and input in the bid/no-bid decision
Knowledge requirements	Technical knowledge (command of technology)	Mostly technical but some business knowledge	Mostly business but some technical knowledge (understanding of technology)
Customer expectations	Deliverables	Deliverables	Business solutions
Definition of success	Meeting the triple constraint	Meeting the triple constraint	Multiple success criteria (both project and business success)

As mentioned in the previous illustration, it was a common belief among executive management that project managers possessed little to no actual business knowledge; and, as a result of that belief, they were brought on board either after contract award or at the end of the initiation phase of the project. Project managers were perceived as "execution" specialists.

As project managers developed their business skills, their input was invited and welcomed during proposal preparation activities. This was particularly true of project managers who demonstrated strong writing skills. Many companies that relied on a competitive bidding process utilized proposal managers. Project managers who were between assignments would often be assigned to proposal managers during competitive bidding activities to prepare various sections of the proposal under the direction of the proposal manager. This provided an opportunity for the project manager to gain additional knowledge about the proposal process and the interrelationships between business entities, and to develop a broader knowledge about the business in general.

As project managers became more knowledgeable about the business, they began to participate in feasibility studies, cost-benefit analyses, and in the bid or no-bid decision processes. Gradually, many companies began to assign project managers the responsibility for developing project business cases and preparation of the entire proposal. Writing skills became an important prerequisite for assignment as a project manager, which was then followed by customer interface, presentation skills, and negotiation skills.

	Historical View	1990	Today
Project manager's role and responsibility	Monitor and control during execution	Planning for project execution	Strategy development and project selection input
When brought on board	After contract award or at end of initiation	During proposal preparation	During concept development and input in the bid/no-bid decision
Knowledge requirements	Technical knowledge (command of technology)	Mostly technical but some business knowledge	Mostly business but some technical knowledge (understanding of technology)
Customer expectations	Deliverables	Deliverables	Business solutions
Definition of success	Meeting the triple constraint	Meeting the triple constraint	Multiple success criteria (both project and business success)

For many years, engineers were assigned as project managers due to their technical knowledge and many had advanced degrees in the engineering profession. This concept was customer driven. Customers required project managers to possess detailed knowledge of the product and a command of technology rather than an understanding of how teams functioned or how to integrate and coordinate deliverables.

But as projects grew in both size and complexity, it became obvious that project managers would be challenged to possess a command of technology in all aspects of the project while managing the day-to-day activities of the project. Also, as project managers spent more time performing project management functions, such as monitoring and assessing project performance, they had less time available to remain technically proficient in their own functional discipline.

As expected, in today's project environment, many project managers have more of an understanding of technology rather than a command of technology. At the same time, the project manager's knowledge of business operations has been growing considerably to the point where project managers are trusted to make major business decisions. Project managers are now viewed as business managers rather than as pure project managers or task managers.

	Historical View	1990	Today
Project manager's role and responsibility	Monitor and control during execution	Planning for project execution	Strategy development and project selection input
When brought on board	After contract award or at end of initiation	During proposal preparation	During concept development and input in the bid/no-bid decision
Knowledge requirements	Technical knowledge (command of technology)	Mostly technical but some business knowledge	Mostly business but some technical knowledge (understanding of technology)
Customer expectations	Deliverables	Deliverables	Business solutions
Definition of success	Meeting the triple constraint	Meeting the triple constraint	Multiple success criteria (both project and business success)

Historically, customers paid contractors for their ability to produce a deliverable. If a second deliverable was required, there was no guarantee that the first contractor would receive the follow-on work. The customers were interested in the end result and not the means to achieve the end result. Some partnerships existed, but the customer, for example, the Department of Defense, wanted to keep as many suppliers as possible involved, thus stimulating competition. While the government's intentions seemed correct at that time, it was at the expense of long-term partnership agreements.

This approach worked well as long as there appeared to be an infinite number of customers and a large contingent of qualified suppliers. For many industries today, the customer base and supplier base have diminished. As such, business owners are seeking out suppliers that can provide their company with long-term business solutions, where the methodology for providing the business solution is included in the project management systems. Contractors, however, are willing to hone their project management skills and provide solutions to the customer's business needs, but in exchange they want to be treated as long-term partners. Project management has therefore matured into a vehicle to obtain long-term strategic business partnerships.

	Historical View	1990	Today
Project manager's role and responsibility	Monitor and control during execution	Planning for project execution	Strategy development and project selection input
When brought on board	After contract award or at end of initiation	During proposal preparation	During concept development and input in the bid/no-bid decision
Knowledge requirements	Technical knowledge (command of technology)	Mostly technical but some business knowledge	Mostly business but some technical knowledge (understanding of technology)
Customer expectations	Deliverables	Deliverables	Business solutions
Definition of success	Meeting the triple constraint	Meeting the triple constraint	Multiple success criteria (both project and business success)

B ecause project managers historically had limited knowledge of the business, the standard definition of success was defined by what is known as the triple constraints: on time, within planned cost, and within quality requirements (or scope/performance). But as the project manager became more knowledgeable about the overall business, the definition of success was modified to contain a business component such as return on investment (ROI) or market share.

While some projects are needed to support cash flow simply to keep the people employed, far too many projects lacked a viable business purpose. Valuable corporate resources were tied up dealing with customer busy work rather than value-added corporate opportunities.

To solve this problem, companies began modifying the definition of project success to include a business component. The existence of the business component made life a little easier during project selection, prioritization, and the portfolio management of projects. This topic will be discussed in more detail in several of the illustrations that follow. Examples will be provided on business components for project success.

	Historical View	1990	Today
Program vs. Project success	Project success is critical	Program success is critical	Project and program success must be integrated
Project management limitations	Company project management	National project management	Global project management is essential for the future
Portfolio management	Handled in secret entirely at the executive levels	Mostly at executive level but some middle management	Heavy involvement by project managers and the project management office

As mentioned in the previous illustration, project success was traditionally measured by performing the work within the triple constraint. This may be regarded as a reasonable definition for success if the project were designed simply to develop a product. What if the product is part of a larger effort such as a program where the product must be marketed and sold as part of a strategic effort? If the project was completed within the parameters of the triple constraint and the company discovers that there is little demand for the product, was the original project a success or a failure?

Historically, companies differentiated between project success and program (or business) success. It was possible to have a successful project but the program could be regarded as a failure. Today, the definition of success has a business component such as ROI, the ability to sell the product, and the ability to satisfy a marketplace need. Slowly but surely, we are including in the definition of success a business component that will align each project with corporate strategic objectives. This will encourage companies to develop a common definition of success that will be equally acceptable at the project and program levels. In the future, project and program success can be expected to be connected and interdependent.

	Historical View	1990	Today
Program vs. Project success	Project success is critical	Program success is critical	Project and program success must be integrated
Project management limitations	Company project management	National project management	Global project management is essential for the future
Portfolio management	Handled in secret entirely at the executive levels	Mostly at executive level but some middle management	Heavy involvement by project managers and the project management office

Historically, project management had some significant limitations regarding its use. Within many companies it was common for project management to be restricted for use in departments that were project driven or project based, such as information technology. All other departments were basically free to use their own approaches or methods to meet their objectives. There was no consistency across the enterprise. Only those departments that were considered project based were using a formal project management methodology.

Each division of a company could decide independently if they wanted to use a project management methodology. The challenge, simply stated, is that it was almost impossible for an entire organization to agree about how to implement project management, let alone accept a common EPM methodology for use by all business components.

Today, project management has become a factor in all levels and divisions within many companies, and this includes the global business environment. Some companies consider this globalization of project management methodology as essential for the survival of the firm. Multinational firms are now managing all projects with a single EPM methodology. The methodology is used to plan and execute projects for all customers, all products, and for the entire product or project life cycle.

	Historical View	1990	Today
Program vs. Project success	Project success is critical	Program success is critical	Project and program success must be integrated
Project management limitations	Company project management	National project management	Global project management is essential for the future
Portfolio management	Handled in secret entirely at the executive levels	Mostly at executive level but some middle management	Heavy involvement by project managers and the project management office

As previously stated, it was believed, historically, that project managers had limited knowledge of business practices and, as such, did not participate in the selection of projects or the management of project portfolios. Portfolio management was viewed as an executive-level function, often done behind closed doors and with little information communicated about the process and the decisions made. Sometimes middle management would be brought on board because they possessed the technical knowledge to assess the risks in each project.

Today, companies are using project management offices, staffed with experienced project managers, to support portfolio management, continuous improvement efforts, and capacity planning activities. Companies are finally recognizing the benefits of having project managers who possess business knowledge.

Some of the more advanced project management offices, primarily in medium-sized to large firms, are establishing internal certification programs for their project managers that are associated with their business processes. Once again, the intent is to expand the capabilities of the project manager to develop a combination of business manager and project manager.

Postulate#1

It doesn't matter whether you execute a project extremely well or extremely poorly if you are working on the wrong project.

The seven postulates show the necessity for the project manager to understand the business and to have a business component as part of the definition of success.

In Postulate #1, we can see what happens when management makes poor decisions during project selection, establishment of a project portfolio, and when managing project portfolios. We end up working on the wrong project or projects. What is unfortunate about this scenario is that we can produce the deliverable that was requested and planned but:

- There is no market for the product.

- The product cannot be manufactured as engineered.

- The assumptions were not validated or may have changed.

- The marketplace and demand may have changed.

- Valuable resources were wasted on the wrong project.

- Stakeholders may be displeased with management's performance.

- The project selection and portfolio management process is flawed.

- Organizational morale has diminished.

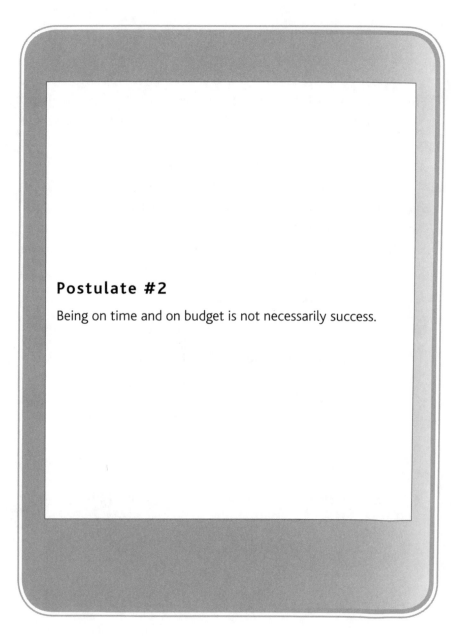

Postulate #2

Being on time and on budget is not necessarily success.

Postulate #2 is the corollary to Postulate #1. Completing a project on time and on budget:

- Does not guarantee a satisfied client/customer
- Does not guarantee that the customer will accept the product/ service
- Does not guarantee that performance expectations will be met
- Does not guarantee that value exists in the deliverable
- Does not guarantee marketplace acceptance
- Does not guarantee follow-on work
- Does not guarantee success

Postulate #3

Completing a project within the triple constraint does not guarantee that the necessary business value will be there at project completion.

Postulate #3 focuses on value. Simply because the deliverable is provided according to a set of constraints is no guarantee that the client will perceive value in the deliverable. The value of a product or service within the context of project management means the relationship between the customer's expectations of product quality and product usefulness, short and long term, to the actual amount paid for it. It is often expressed as the equation:

$$Value = Benefits/Price$$

or

$$Value = Quality\ received/Expectations$$

There are also parallels between cultural expectations and customer expectations when considering value.

The ultimate objective of all projects should be to produce a deliverable that meets expectations and achieves the desired value. While we always seem to emphasize the importance of the triple constraint when defining the project, we spend very little time in defining the value characteristics that we expect in the final deliverable.

The value component or definition must be a joint agreement between the customer and the contractor (buyer/seller) during the initiation stage of the project. Also, in the ideal situation, **the definition of value is aligned with the strategic objectives of both the customer and the contractor.**

Postulate #4

Having mature project management practices, including an EPM methodology, does not guarantee that business value will be there at project completion.

Postulate #4 focuses on success and how it cannot be guaranteed. Most companies today have some type of project management methodology in place. Unfortunately, all too often, there is a mistaken belief that the methodology will guarantee project success. It should be understood that methodologies:

- Cannot guarantee success.

- Cannot guarantee value in the deliverable.

- Cannot guarantee that the time constraint will be adhered to.

- Cannot guarantee that the quality constraint will be met.

- Cannot guarantee any level of performance.

- Are not a substitute for effective planning.

- Are not the ultimate panacea to cure all project ills.

- Are not a replacement for effective management.

- Will not guarantee effective human behavior.

Methodologies can improve the chances for success but the use of a methodology cannot guarantee the successful delivery of a desired product or service. Methodologies are tools and, as such, do not manage projects. Projects are managed by people and, likewise, tools are managed by people. Methodologies do not replace the people component in project management. They are designed to enhance the performance of people.

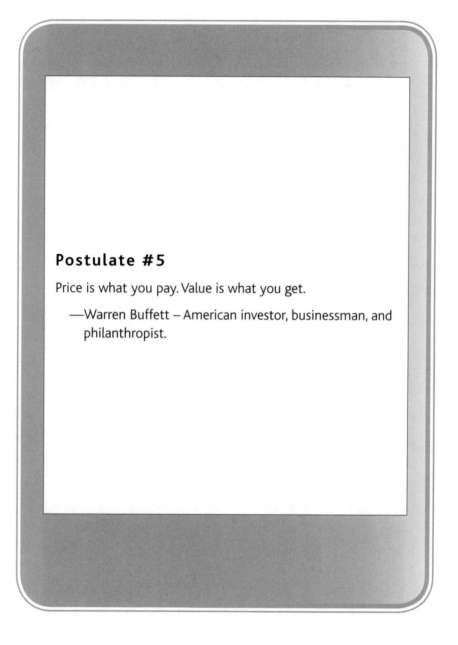

Postulate #5

Price is what you pay. Value is what you get.

  —Warren Buffett – American investor, businessman, and
    philanthropist.

Postulate #5 focuses on the quote "Price is what you pay. Value is what you get." This quote from Warren Buffet emphasizes the difference between price and perceived value. Most people believe that customers pay for deliverables. This is not necessarily true. Customers pay for the value they expect to receive from the deliverable. If the deliverable has not achieved value or has limited value, the result is a dissatisfied customer.

Some people believe that a customer's greatest interest is quality. In other words, "Quality comes first!" While that may seem to be true on the surface, the customer generally does not expect to pay an extraordinary amount of money just for high quality. Quality is just one component in the value equation. Value is significantly more than just quality.

For a firm to deliver value to its customers, it must consider what is known as the "total product offering." This includes the reputation of the organization, staff representation and support, product benefits, technological characteristics, ease of use, reliability, and brand recognition as compared to competitors' market offerings and prices. Value could also be defined as the relationship of company's market offerings to those of its competitors.

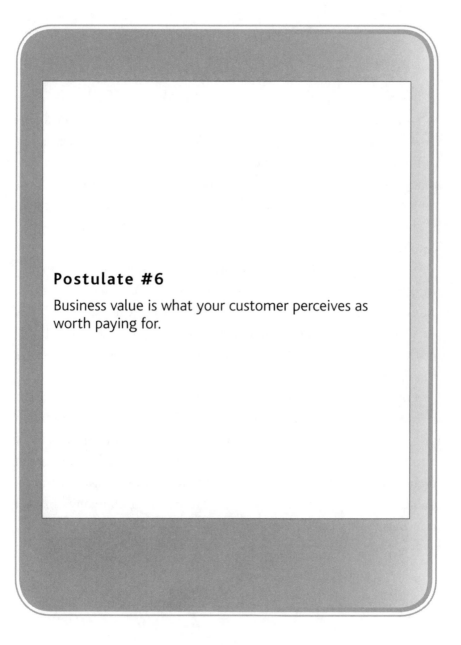

Postulate #6

Business value is what your customer perceives as worth paying for.

Postulate #6 focuses on business value and that business value is what your customer perceives as worth paying for. When customers agree to a contract with a contractor/supplier for a deliverable, the customer is actually looking for the value in the deliverable. The customer's definition of success is "value achieved."

Unfortunately, unpleasant things can happen when the project manager's definition of success is the achievement of the deliverable (and possibly the triple constraint) and the customer's definition of success is value. This is particularly true when customers want value, and you, as the contractor, focus on the profit margins of your projects. This approach can lead to severe conflict between buyer and seller and possibly litigation.

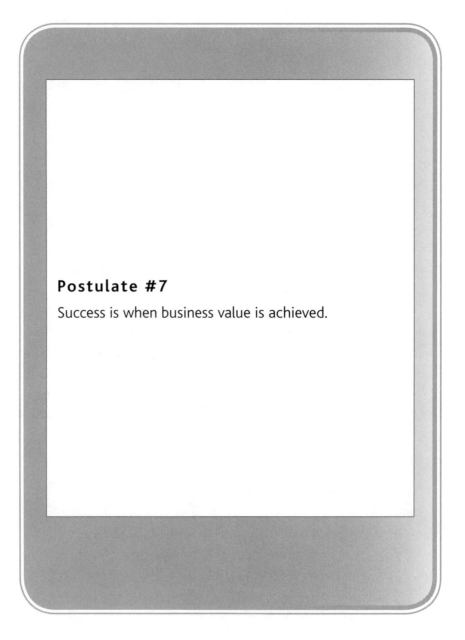

Postulate #7

Success is when business value is achieved.

Postulate #7 is a summation of Postulates #1 through #6. Perhaps the standard definition of success using the triple constraint should be modified to include a business component such as value, or even be replaced by a more specific definition of value. Examples of value include:

- Availability—the product will perform when called upon to do so

- Operability—the degree to which a product can be operated safely

- Reliability—the product will perform without failure for a set period of time

- Maintainability—the product can be retained in or restored to an acceptable level of performance

- Social acceptability—the degree of conflict between the product and the values of society is minimized (safety, environmental issues)

- Cultural acceptance—the degree to which a product challenges or conflicts with customs and beliefs of the intended market population

RECOGNIZING THE NEED FOR CHANGE

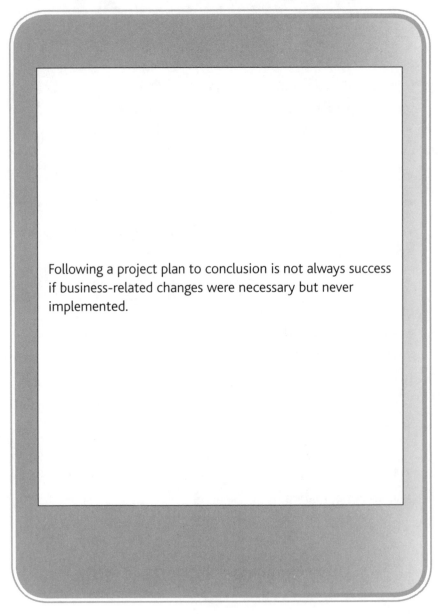

Following a project plan to conclusion is not always success if business-related changes were necessary but never implemented.

Sometimes the value of a project can change over time and the project manager may not recognize that these changes have occurred. Failure to establish value expectations or lack of value in a deliverable can result from:

- Market unpredictability

- Market demand changed

- Changing constraints and assumptions

- Technology advances or inability to achieve functionality

- Critical resources were either not available or resources lacked the necessary skills

The value component may indicate that some projects should be canceled. The earlier the project is canceled, the sooner the resources can be assigned to projects that have a higher perceived value and probability of success. Unfortunately, early warning signs are not always present to indicate that the value will not be achieved.

Earned value measurement techniques focus on tracking the triple constraint. Earned value measurement techniques should also track the validity of the project assumptions and forecast the value or benefits at completion, if possible.

CHANGING OUR DEFINITION OF PROJECT SUCCESS

CHANGING TIMES

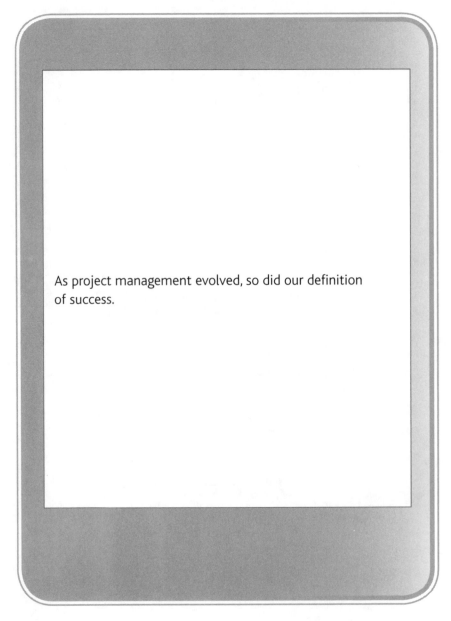

As project management evolved, so did our definition of success.

The definition of success continues to evolve as project management matures. Over a four-year period, one telecom company changed its definition of success as follows:

- Year 1: Completing a project

- Year 2: Completing a project on time

- Year 3: Completing a project on time and with quality

- Year 4: Meeting products expectations according to a customer-agreed-upon solution

Today, both customers and contractors must agree upon the definition of success early in the planning process. This is usually accomplished in the engagement meeting between the customer and the contractor. In this meeting the expectations of the customer are defined through discussions about reliability, safety, warranties, fitness for use, and the timeliness of communication.

As the definition of success changes, so must the definition of value.

NOT MEETING THE TRIPLE CONSTRAINT

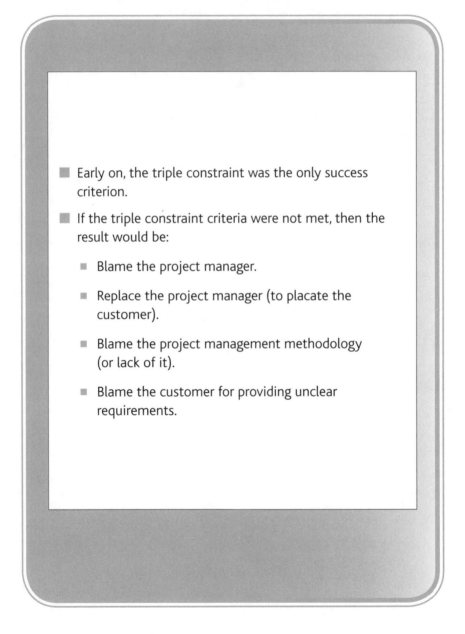

- Early on, the triple constraint was the only success criterion.
- If the triple constraint criteria were not met, then the result would be:

  - Blame the project manager.

  - Replace the project manager (to placate the customer).

  - Blame the project management methodology (or lack of it).

  - Blame the customer for providing unclear requirements.

Historically, the triple constraint was considered to be the only truly measurable success criterion for a project. If each element of the triple constraint could not be met, blame was usually placed on the project team and the project manager when, in fact, the real culprit may have been **poor business decision making.** Project managers were either fired or removed from the project. To appease the customer, the contractor might use the following explanation and strategy:

> One of our best project managers is now available, and he/she will be assuming responsibility for the project. We'll be back on track before you know it!

The bad news is that you never really get back on track. Typical scenarios such as the following may develop:

- The project is shut down until replanning takes place.

- The new project manager wants to introduce a new or fresh plan that can be considered his or her own, and disregards everything accomplished thus far.

- Critical resources look for opportunities to desert a sinking ship.

- When the project finally resumes, it has slipped much further behind than before.

- The expected value is not achieved.

DEFINING PROJECT AND PROGRAM SUCCESS

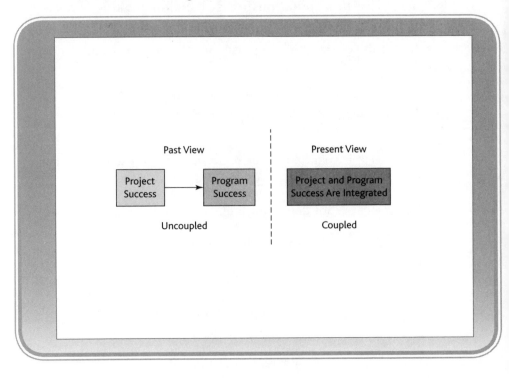

Historically, the definitions of project success and program success were uncoupled. A successfully completed project was no guarantee that program success would be achieved. It was common practice to have two different definitions, with project success defined more so in technical terms and program success defined in business terms. Sometimes, program success would be defined only after the projects supporting it had been completed.

Today, companies are integrating project and program success into one definition: meeting the business results and business expectations (i.e., value) as well as the triple constraint and other success criteria. The business component can be defined in financial terms as:

- Return on investment (ROI)

- Payback period

- Net present value

- Internal rate of return (IRR)

- Market share

- Gross sales

- Customer base

REDEFINING THE TRIPLE CONSTRAINT
SUCCESS CRITERIA

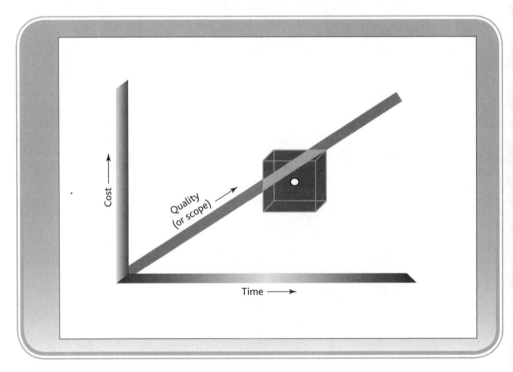

The triple constraint of time, cost, and performance is the circle in the middle of the cube. But, in reality, the cube is the actual definition of success. As an example:

- Completing a project $10,000 over budget could still be viewed as success based on the size of the original budget and the value that was achieved.

- Completing a project two weeks late could still be viewed as success based on the duration of the project and the willingness of the customer to allow for a slippage if they perceive that the final product or deliverable will meet their needs.

- Providing a deliverable that meets only 92 percent of the specification requirements could still be viewed as a success in the eyes of the customer if the functionality addresses their most critical requirements.

Actually, the definition of success is completing the project within the cube rather than the triple constraint. Unfortunately, most executives do not define or provide the project manager with the boundaries of the cube.

DEFINITION OF SUCCESS

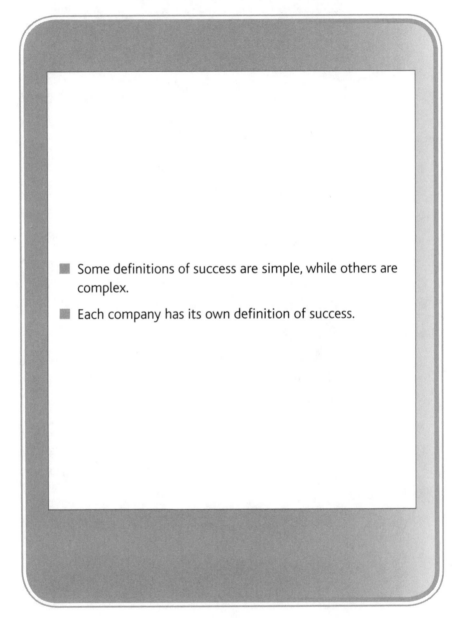

- Some definitions of success are simple, while others are complex.
- Each company has its own definition of success.

Each company or organization has its own definition of success. The definition depends on the culture of the organization and its "mission critical issues." Below are two definitions of success*:

ORANGE SWITZERLAND:

- The delivery of the "product" within time, cost, and quality characteristics

- The successful management of changes during the project life cycle

- The management of the project team

- The success of the product against defined criteria and targets established during the project initiation phase (adoption rates, ROI, etc.)

NORTEL NETWORKS:

- Nortel defines project success based on schedule, cost, and quality measurements, as mutually agreed to with the customer, the project team, and key stakeholders. Real project success, however, is ultimately measured by customer satisfaction.

The key to adding value is that the definition of success must be agreed upon between the customer and the contractor.

---

*Source: Harold Kerzner, *Project Management Best Practices: Achieving Global Excellence, 1st ed.* Hoboken , NJ: John Wiley & Sons, © 2006. pp. 23–26.

Chapter

3

THE IMPORTANCE OF VALUE

SUCCESS

Success is not necessarily achieved by completing the project within the triple constraint. Success is when the planned business value is achieved within the imposed constraints and assumptions.

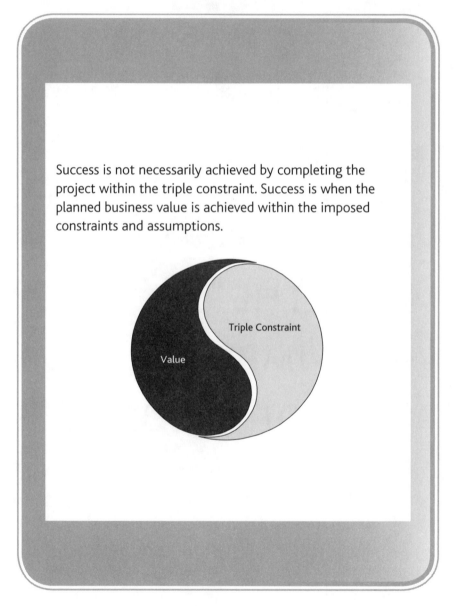

As shown, value must be somehow included in the definition of success. Unfortunately, this is easier said than done because value is a set of beliefs related to what is important to the customer and the stakeholders associated with the project. Value may actually be a perception in many cases. To achieve perceived value and ultimately project success, the contractor must be able to "see" value through the eyes of the customer.

TYPES OF VALUE

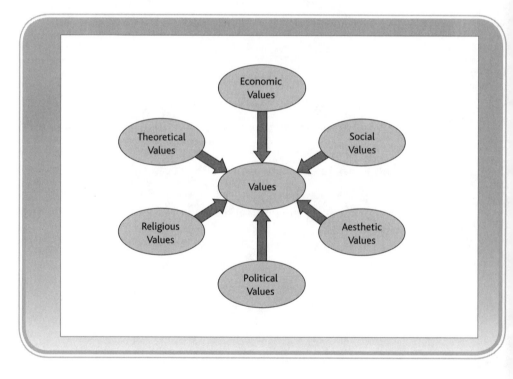

As seen in the illustration on the previous page, there are many types of value. Because professional responsibility is now an integral part of project management, we could argue that all of these types of value apply to project management, including:

- Adhering to ethical standards

- Receiving or providing gifts

- Adhering to security and confidentiality requirements

- Truthfully reporting information

- Willingness to identify violations

- Recognizing and accepting diversity

- Managing customer/contractor intellectual property

- Maintaining professional integrity

- Understanding global religious, cultural, and ethical beliefs

For this module, project management value entails the attributes of your enterprise project management (EPM) system, appearing in the form of goods or services, for which your customers are willing to pay. The project, using the EPM system, must produce distinct products as services with features and characteristics that are valued by the customers.

RETURN ON INVESTMENT (ROI)

ROI is not the only measurement of business-related value.

ROI may be the main definition of value for an enterprise or for a project manager, but it is certainly not the only definition of value. The customer may have a completely different definition or set of definitions of value. The customer's definition of value may be associated with quality, compliance with regulations, ease of use, flexibility, adaptability, and other such items. Project managers must seek to define and understand the customer's definition of value. This topic will be discussed in more depth later in the text.

TYPES OF BUSINESS VALUES

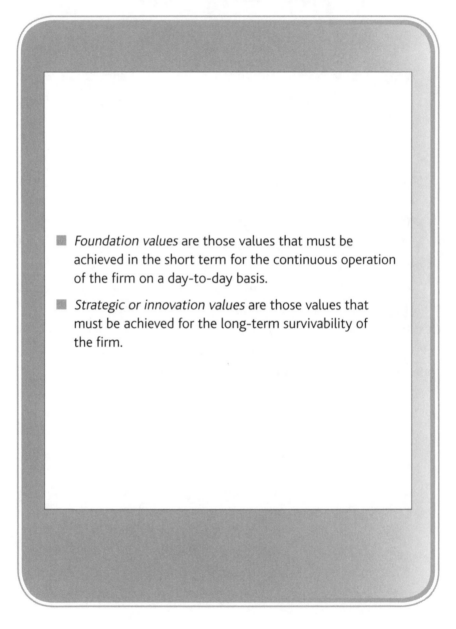

■ *Foundation values* are those values that must be achieved in the short term for the continuous operation of the firm on a day-to-day basis.

■ *Strategic or innovation values* are those values that must be achieved for the long-term survivability of the firm.

Obviously, there are several ways by which business or economic value can be classified. We will use two classifications, foundation values and strategic or innovation values.

Foundation values affect the culture of the company in the way that people work together. This is dependent on terms such as *teamwork, communications, cooperation,* and *trust.* A good EPM methodology can make it easier for the foundation values to exist. By streamlining forms, guidelines, templates, and checklists, people are more willing to accept the EPM methodology. Customers can have a direct influence on foundation values.

Strategic or *innovation values* usually come after the foundation values are achieved. Strategic values are most often internal values to the firm and can include:

- Establishing and maintaining a degree of market share

- Brand recognition

- Satisfying government regulations such as the Occupational Safety and Health Administration (OSHA) and the Environmental Protection Agency (EPA)

- Maintaining regulatory agency relations

- Maintaining ethical conduct

- Maintaining the corporate reputation and image

- Development of intellectual property

- Maintaining a leadership position in the field

CHANGING VALUES

Moving Away From: (Ineffective Values)	Moving Toward: (Effective Values)
Mistrust	Trust
Job descriptions	Competency models
Power and authority	Teamwork
Internal focus	Stakeholder focus
Security	Taking risks
Conformity	Innovation
Predictability	Flexibility
Internal competition	Internal collaboration

(Adapted from Ken Hultman and Bill Gellerman, *Balancing Individual and Organizational Values.* Jossey-Bass/Pfeiffer, a Wiley Company, © 2002, pp.105–106.)

The next several illustrations show examples of the way values are changing. Most of these changes appear to be in the foundation values. Trust is the reliance on the integrity, strength, knowledge, ability, and decisions of others. Project management generally requires that employees from various functional groups meet, form teams, work together as a group toward a common objective, and then return to their functional group. It is possible that these employees will never interface with these same team members again.

Trust between individuals usually takes a great deal of time to develop. But in project management, time is a constraint rather than a luxury. People assigned to project teams must trust their colleagues to do the right thing and make the right decisions. Mistrust will elongate projects by forcing many decisions to be questioned or reevaluated.

Project managers who mistrust the team members usually try to revalidate all of the numbers and equations that may have gone into the final decision. Not only does this have the potential to increase the duration of the project; it also has a severe negative impact on morale and creates a feeling of inferiority or lack of confidence and a reduction in productivity. When project team members are treated in this manner, morale suffers and team members may avoid working for the project manager on future assignments.

Moving Away From: (Ineffective Values)	Moving Toward: (Effective Values)
Mistrust	Trust
Job descriptions	Competency models
Power and authority	Teamwork
Internal focus	Stakeholder focus
Security	Taking risks
Conformity	Innovation
Predictability	Flexibility
Internal competition	Internal collaboration

(Adapted from Ken Hultman and Bill Gellerman, *Balancing Individual and Organizational Values*. Jossey-Bass/Pfeiffer, a Wiley Company, © 2002, pp.105–106.)

Writing a job description for a project manager is easy. Just describe the specific duties and responsibilities associated with planning, organizing, and integrating deliverables of the project. Writing job descriptions that differentiate between pay grades of project managers is much more difficult. This is why companies are using alternate methods for developing job descriptions when positioning project management as a career path in the company.

Twenty years ago, companies prepared job descriptions for project managers by explaining only the roles and responsibilities. Unfortunately, job descriptions were usually abbreviated and provided little guidance about what was required for advancement and salary increases. The success of the project was the underlying factor. Ten years ago, companies were emphasizing the importance of job descriptions, but they were supported by training, coursework, and education, which was often mandatory.

Today, job descriptions are being replaced with **project management core competency models,** which emphasize the skills needed to be an effective project manager. Training programs are being developed and implemented to support core competency models.

Moving Away From: (Ineffective Values)	Moving Toward: (Effective Values)
Mistrust	Trust
Job descriptions	Competency models
Power and authority	Teamwork
Internal focus	Stakeholder focus
Security	Taking risks
Conformity	Innovation
Predictability	Flexibility
Internal competition	Internal collaboration

(Adapted from Ken Hultman and Bill Gellerman, *Balancing Individual and Organizational Values.* Jossey-Bass/Pfeiffer, a Wiley Company, © 2002, pp.105–106.)

Traditionally, project managers negotiated and sometimes fought for the maximum amount of authority (and power) they could attain. Executives, however, felt threatened by the potential power and influence of the project manager and feared that the decentralization of authority and decision making would bring chaos to the organization and result in havoc and conflict within projects. There was a belief that a loss of control would be experienced at the executive levels.

Today, project managers have come to the realization that **very little authority and power is needed to be effective as a project manager.** Today, the *value attribute* of a project manager appears to be teamwork. Historically, whatever authority the project manager had came from either job descriptions or project charters. Providing the project manager with a certain level of authority was deemed necessary because of the relationship and potential for conflict or mistrust between the project manager and the team members.

When trust exists, the focus is on teamwork and collaboration. Leadership in project management is centered around team performance rather than just the efforts of the project manager.

Moving Away From: (Ineffective Values)	Moving Toward: (Effective Values)
Mistrust	Trust
Job descriptions	Competency models
Power and authority	Teamwork
Internal focus	Stakeholder focus
Security	Taking risks
Conformity	Innovation
Predictability	Flexibility
Internal competition	Internal collaboration

(Adapted from Ken Hultman and Bill Gellerman, *Balancing Individual and Organizational Values.* Jossey-Bass/Pfeiffer, a Wiley Company, © 2002, pp.105–106.)

Historically, the project manager's definition of success was focused internally within the company and was usually defined in terms of profitability, the triple constraint, and individual ability. Today, the focus appears to have shifted to the stakeholders, especially the customer, rather than on profit and achievement of company objectives. The belief that satisfied customers result in achievement of organizational goals is now becoming the prevalent management perspective.

Marketing and sales people today are emphasizing "engagement project management" and total business solutions. The continuous change in the global business environment and the greater competitive nature of business has resulted in the fact that we no longer have an infinite supply of customers. Therefore, a successful initial engagement with the client is essential. Contractors and suppliers must sell the customer on project management methodology, their ability to capture best practices and share them with the client, and the ability to provide the customers with a long-term solution to their business needs.

In exchange for this, we want the client to treat us as a long-term strategic partner rather than just another contractor. In other words, we view our project management systems and processes as the vehicles to maintain long-term stakeholder partnerships.

Moving Away From: (Ineffective Values)	Moving Toward: (Effective Values)
Mistrust	Trust
Job descriptions	Competency models
Power and authority	Teamwork
Internal focus	Stakeholder focus
Security	Taking risks
Conformity	Innovation
Predictability	Flexibility
Internal competition	Internal collaboration

(Adapted from Ken Hultman and Bill Gellerman, *Balancing Individual and Organizational Values.* Jossey-Bass/Pfeiffer, a Wiley Company, © 2002, pp.105–106.)

Historically, project managers walked the straight-and-narrow path and avoided taking undue risks on projects. This approach was influenced greatly by the risk-taking culture of the organization. While this appeared as a reasonable and acceptable approach, it limited many growth opportunities. Today, project managers emphasize risk management and value their abilities to assess and determine which risks are acceptable. Executives expect project managers to be proficient in risk management and understand the risks associated with the particular industry, overall business risks, and how project risks may impact the operations of the company.

Planning focuses on history and lessons learned. Risk management forces the project manager to look ahead and "anticipate" both favorable and unfavorable events that can impact the success of the project. Risk management allows us to make maximum usage of the opportunities to improve the value of the deliverable. Effective risk management cannot be accomplished without a thorough knowledge of the business, the industry, and the environmental factors that affect an organization.

Moving Away From: (Ineffective Values)	Moving Toward: (Effective Values)
Mistrust	Trust
Job descriptions	Competency models
Power and authority	Teamwork
Internal focus	Stakeholder focus
Security	Taking risks
Conformity	Innovation
Predictability	Flexibility
Internal competition	Internal collaboration

(Adapted from Ken Hultman and Bill Gellerman, *Balancing Individual and Organizational Values*. Jossey-Bass/Pfeiffer, a Wiley Company, © 2002, pp.105–106.)

Previously, we discussed security which is often achieved through conformity. Project management methodologies are based on conformity, consistency, and standardization to minimize the possibility of repeating mistakes. Lessons learned are captured during the project life cycle and shared with the organization to improve the effectiveness of the processes and ensure conformity.

Innovation thrives on the taking of risks. Without taking risks, a company will miss many opportunities and become a follower rather than a leader in their particular industry. Effective and well-developed project management methodologies allow for some degree of innovation.

Moving Away From: (Ineffective Values)	Moving Toward: (Effective Values)
Mistrust	Trust
Job descriptions	Competency models
Power and authority	Teamwork
Internal focus	Stakeholder focus
Security	Taking risks
Conformity	Innovation
Predictability	Flexibility
Internal competition	Internal collaboration

(Adapted from Ken Hultman and Bill Gellerman, *Balancing Individual and Organizational Values*. Jossey-Bass/Pfeiffer, a Wiley Company, © 2002, pp.105–106.)

Conformity breeds predictability, but this is not always desirable. For innovation to exist, project managers must develop some degree of flexibility and the ability to see opportunity in what others might view as failure. Gene Kranz, flight director for Apollo 13, saw many opportunities to save the astronauts, while others where convinced that the outcome would be NASA's greatest disaster. Many new innovations come from the exploration of unusual results of a research and development (R&D) test.

One of the advantages of having and using an EPM methodology is that it promotes standardization, consistency, and conformity. This can be done with policies and procedures. Unfortunately, policies and procedures do not generally allow for flexibility, and flexibility is needed for risk taking and innovation. This is why most EPM systems are designed around forms, guidelines, templates, and checklists.

Moving Away From: (Ineffective Values)	Moving Toward: (Effective Values)
Mistrust	Trust
Job descriptions	Competency models
Power and authority	Teamwork
Internal focus	Stakeholder focus
Security	Taking risks
Conformity	Innovation
Predictability	Flexibility
Internal competition	Internal collaboration

(Adapted from Ken Hultman and Bill Gellerman, *Balancing Individual and Organizational Values.* Jossey-Bass/Pfeiffer, a Wiley Company, © 2002, pp.105–106.)

For decades, project managers would compete with other project managers within the same firm for the resources they needed. Internal competition existed to the point where one project manager would refuse to help another project manager unless directed to do so by the project sponsors. This type of internal competition actually diminished the value-adding capabilities of the organization.

Today, the focus has intensified around internal collaboration, with competition being viewed as those factors that are external to the company that may prevent the achievement of business objectives. With internal competition, project managers make decisions in the best interest of their project and with little regard for the ongoing business of the firm. **By focusing on internal collaboration, project managers tend to make more effective business decisions.** This requires the project manager to have a greater knowledge of the organization's business goals and initiatives.

Internal collaboration allows the project managers to negotiate for the *needed* resources rather than the best resources. It may not be desirable for the best resources to be assigned to a low-priority project. Also, with internal collaboration, project managers are more willing to sacrifice their project by releasing resources or even recommending cancellation of the project so that the resources can be used on those projects that offer a greater opportunity to provide value to the organization or to the customer.

Moving Away From: (Ineffective Values)	Moving Toward: (Effective Values)
Reactive management	Proactive management
Formality	Informality
Bureaucracy	Boundaryless
Traditional education	Lifelong education
Hierarchical leadership	Multidirectional leadership
Tactical thinking	Strategic thinking
Compliance	Commitment
Meeting standards	Continuous improvements

(Adapted from Ken Hultman and Bill Gellerman, *Balancing Individual and Organizational Values.* Jossey-Bass/Pfeiffer, a Wiley Company, © 2002, pp.105–106.)

As stated previously, in the early years of what is now known as formal project management, most project managers were engineers with advanced degrees and significant technical knowledge and experience. These engineers generally believed that whatever plan they laid out, because of their expertise and knowledge of the technology, would work as predicted, and there was very little need for contingency plans. Hence, reactive project management became the norm. Project managers waited for a crisis to occur before considering alternatives.

Today, because risk management has taken on critical importance in the business and government communities, project managers recognize **proactive management as adding value**. Project managers now prepare contingency plans to ensure that if a crisis occurs, specific actions and steps are already in place and may be executed to remedy the situation. Without proactive management, valuable time may be lost and critical resources may be reassigned to other projects.

Proactive management grew in importance as the importance of risk management grew. Risk management forced project managers to look ahead and anticipate risk events. The concept of identifying risk triggers (symptoms of risks) also supported proactive planning.

Moving Away From: (Ineffective Values)	Moving Toward: (Effective Values)
Reactive management	Proactive management
Formality	Informality
Bureaucracy	Boundaryless
Traditional education	Lifelong education
Hierarchical leadership	Multidirectional leadership
Tactical thinking	Strategic thinking
Compliance	Commitment
Meeting standards	Continuous improvements

(Adapted from Ken Hultman and Bill Gellerman, *Balancing Individual and Organizational Values.* Jossey-Bass/Pfeiffer, a Wiley Company, © 2002, pp.105–106.)

Formalized project management thrives on policies and procedures accompanied by mountains of paperwork. Formality usually occurs because of mistrust, inconsistency, and management's desire to have a rigid mechanism in place for projects for control purposes. While there is some merit to formality—namely, conformity, standardization, and control—it does limit flexibility and the opportunity for innovation.

Today, the focus is more toward informality. Informality does have some degree of formality in it, but emphasis is on creating more efficient processes and a "paperless" project management system. Informality gives the project manager tremendous flexibility in how the EPM system will be adapted to a particular customer's needs. Informality also stresses the need for forms, guidelines, templates, and checklists rather than policies and procedures. Informality does not eliminate paperwork entirely, but it does reduce it significantly.

Informality allows the project manager to custom design the project management methodology for a particular customer with the objective of building a long-term business relationship. This is in line with our previous discussion of engagement project management and engagement selling.

Moving Away From: (Ineffective Values)	Moving Toward: (Effective Values)
Reactive management	Proactive management
Formality	Informality
Bureaucracy	Boundaryless
Traditional education	Lifelong education
Hierarchical leadership	Multidirectional leadership
Tactical thinking	Strategic thinking
Compliance	Commitment
Meeting standards	Continuous improvements

(Adapted from Ken Hultman and Bill Gellerman, *Balancing Individual and Organizational Values.* Jossey-Bass/Pfeiffer, a Wiley Company, © 2002, pp.105–106.)

Bureaucracy is a derivative of extremely formalized project management. In a bureaucratic organization, all project decisions must follow a very specific chain of command. Project managers basically assume the role of project monitors rather than performing as coordinator and integrator, and have limited authority to make decisions. Any decisions that are required must be reviewed, verified, and approved through the organization's chain of command.

Project management is a methodology designed to allow work to flow in a multidirectional manner through a company. Project managers should be given the authority and support to communicate directly with the managers or representatives of each business unit or department in the organization to obtain the necessary project information. Forcing all of the information through the highly bureaucratic chain of command slows down project progress and can considerably delay the decision-making process. This could result in lost opportunities, customer dissatisfaction, and additional planning delays.

Bureaucracy is also an expensive way to run an organization. Unnecessary layers of management are superimposed upon both the horizontal and vertical work flows. Team members are not empowered to make decisions for their specific line of business or functional unit. Line management attendance is mandatory at all decision-making team meetings.

Moving Away From: (Ineffective Values)	Moving Toward: (Effective Values)
Reactive management	Proactive management
Formality	Informality
Bureaucracy	Boundaryless
Traditional education	Lifelong education
Hierarchical leadership	Multidirectional leadership
Tactical thinking	Strategic thinking
Compliance	Commitment
Meeting standards	Continuous improvements

(Adapted from Ken Hultman and Bill Gellerman, *Balancing Individual and Organizational Values.* Jossey-Bass/Pfeiffer, a Wiley Company, © 2002, pp.105–106.)

Companies that wish to maximize the organization's ROI in project management do so through formalized education. But simply offering an employee one course on the fundamentals of project management does not achieve a very high return on the investment. **Lifelong education is the key** to long-term growth in project management. Traditionally, project management was viewed as planning, scheduling, and cost control. Today, project management has aligned itself with other management concepts, all requiring education, such as:

- Six Sigma
- International Organization for Standardization (ISO 9000)
- Capability Maturity Model (CMM)
- Organizational Project Management Maturity Model (OPM3)
- Concurrent engineering
- Life-cycle costing
- Configuration management
- Risk management
- Portfolio management
- Capacity planning
- Project offices and centers of excellence
- Global team management
- Virtual teams

Moving Away From: (Ineffective Values)	Moving Toward: (Effective Values)
Reactive management	Proactive management
Formality	Informality
Bureaucracy	Boundaryless
Traditional education	Lifelong education
Hierarchical leadership	Multidirectional leadership
Tactical thinking	Strategic thinking
Compliance	Commitment
Meeting standards	Continuous improvements

(Adapted from Ken Hultman and Bill Gellerman, *Balancing Individual and Organizational Values.* Jossey-Bass/Pfeiffer, a Wiley Company, © 2002, pp.105–106.)

Previously, we discussed the changing management approach from bureaucracy to boundaryless project management and the pursuit of value. Within bureaucracy, all leadership is hierarchical and follows a rigid chain of command. Each employee takes direction from one and only one assigned manager, and employees often work in isolated silos with little knowledge of the overall business operation.

Successful project management is a form of *multidirectional leadership* where information and work flows horizontally and vertically in a simultaneous manner. The good news is that this allows the organization to accomplish more work in less time and with fewer resources without any significant sacrifice in quality. The bad news, or where problems may arise, is that employees must now take direction from their line or functional group manager and the many project managers on whose teams they may be assigned. This could result in some conflict about the availability of resources and the priority of the project.

Multidirectional leadership is very effective if project managers and line managers can work together, collaborate, and coordinate their direction with respect to the employees or project performers and the critical issues of the company. Without this coordination, several projects will experience slippages, customer satisfaction issues, and other severe conflicts that may result in direct involvement by the project sponsor or executive steering committee. In this case, multidirectional leadership may be viewed as a failure.

Moving Away From: (Ineffective Values)	Moving Toward: (Effective Values)
Reactive management	Proactive management
Formality	Informality
Bureaucracy	Boundaryless
Traditional education	Lifelong education
Hierarchical leadership	Multidirectional leadership
Tactical thinking	Strategic thinking
Compliance	Commitment
Meeting standards	Continuous improvements

(Adapted from Ken Hultman and Bill Gellerman, *Balancing Individual and Organizational Values*. Jossey-Bass/Pfeiffer, a Wiley Company, © 2002, pp.105–106.)

As discussed in several illustrations, project managers in the past typically had little business knowledge and were brought into the project after initiation was complete. Project managers were considered to be tacticians rather than strategic thinkers. Today, project managers possess significant knowledge about business processes and a broader view of the enterprise and are expected to make strategic as well as tactical decisions. Project managers in today's environment must think well beyond the end date of the project.

It was a common practice to assign project managers to projects after the business case was developed. They were given the responsibility and were held accountable to execute the project successfully but had no input in the decision processes and factors associated with project selection and anticipated value. Today, because project managers have some degree of business knowledge, they now participate in those activities that require strategic thinking, namely, business case development, portfolio management and selection of projects, project prioritization techniques, and capacity planning.

Strategic project management is also important during mergers and acquisitions. If two companies are to join forces, synergy will not be achieved unless the project management systems of both companies can be effectively combined.

Moving Away From: (Ineffective Values)	Moving Toward: (Effective Values)
Reactive management	Proactive management
Formality	Informality
Bureaucracy	Boundaryless
Traditional education	Lifelong education
Hierarchical leadership	Multidirectional leadership
Tactical thinking	Strategic thinking
Compliance	Commitment
Meeting standards	Continuous improvements

(Adapted from Ken Hultman and Bill Gellerman, *Balancing Individual and Organizational Values*. Jossey-Bass/Pfeiffer, a Wiley Company, © 2002, pp.105–106.)

Compliance is the act or tendency to yield to others either freely or by force. Commitment is the act of willingly agreeing to something such as a set of objectives, a budget, or schedule or plan. **Commitment usually comes with agreement, whereas compliance does not.**

It is very difficult to force people to comply with a date or a budget, especially if they had no input to the decision process. "Lip service" (providing information or feedback that is insincere or basically useless) is a form of compliance and may not be indicative of the actual support needed by the project manager. Compliance can be attained at the expense of teamwork.

During employee staffing activities, line managers sometimes ask the employees whether they would like to be assigned to a given project. If the employee accepts the assignment, it is usually accompanied by a commitment. If the line manager simply forces the employee to work on a given project with no explanation or opportunity for discussion, the result can be compliance but without commitment. In these cases, the quality of the project deliverables may be severely affected.

Moving Away From: (Ineffective Values)	Moving Toward: (Effective Values)
Reactive management	Proactive management
Formality	Informality
Bureaucracy	Boundaryless
Traditional education	Lifelong education
Hierarchical leadership	Multidirectional leadership
Tactical thinking	Strategic thinking
Compliance	Commitment
Meeting standards	Continuous improvements

(Adapted from Ken Hultman and Bill Gellerman, *Balancing Individual and Organizational Values*. Jossey-Bass/Pfeiffer, a Wiley Company, © 2002, pp.105–106.)

Today, capturing, documenting, and utilizing best practices in project management can be viewed as a competitive weapon for an enterprise, especially during negotiations and competitive bidding. When a project is completed, project managers should be expected to collaborate with the project management office to debrief the project team and review the major lessons learned and best practices discovered during project implementation. The documentation of best practices should be considered part of a company's continuous improvement efforts. Without a focus on continuous improvements, the company can falter, fall behind in the market, and eventually fail.

Another problem generally associated with standards is that people have the tendency to stop improving once a standard is achieved. **People should be encouraged to continuously improve rather than just meet the standards.** If standards are used, the bar should be raised whenever the standard is achieved.

Continuous improvement is also competitive weapon. It is an essential component of engagement project management and engagement selling. One of the main functions of a project management office (PMO) or project center of excellence is to support continuous improvement by monitoring organizational project performance and identifying innovative techniques that improve the probability of success in achieving value. These techniques are documented and shared as part of the organization's continuous improvement efforts.

Chapter

THE
STAKEHOLDERS'
VIEW OF VALUE

STAKEHOLDER PERCEPTION

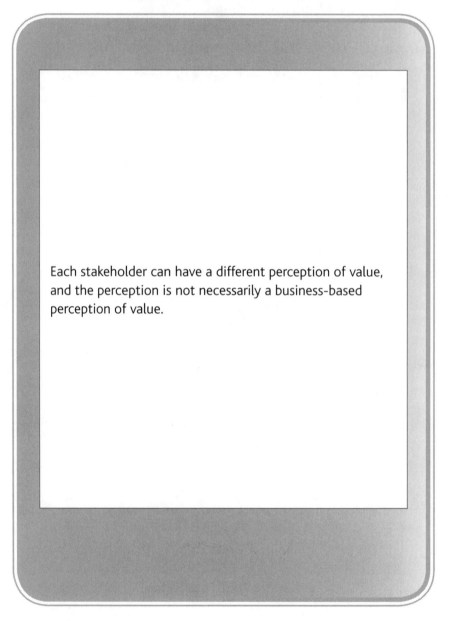

Each stakeholder can have a different perception of value, and the perception is not necessarily a business-based perception of value.

Project managers must deal with a multitude of stakeholders, and each stakeholder will probably have a different perception of value. As an example, when the Tylenol tragedies occurred in the 1980s, Johnson & Johnson had to make decisions based on competing stakeholder demands. The stakeholders included:

- Consumers

- Stockholders

- Lending institutions

- Government agencies

- Corporate management

- Employees

Each stakeholder wanted the Tylenol tragedies problem to be resolved in a manner that protected their perception of value. Some stakeholders viewed value in financial terms such as dividends, while others identified value as providing job security. Others viewed value as providing consumers with safe products and compassion.

Balancing stakeholders needs becomes quite complex when the projects span international borders. Culture, ethics, religion, and government bureaucracy become important considerations.

CLASSIFICATION OF STAKEHOLDERS

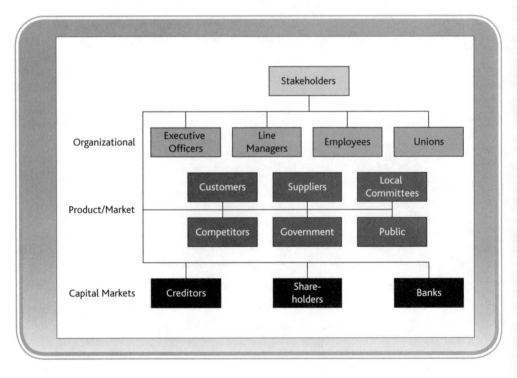

This illustration shows typical classification systems for stakeholders. For simplicity's sake, the stakeholders can be classified as:

- Organizational stakeholders
- Product/market stakeholders
- Capital market stakeholders

There are other classification systems, and each system may be dependent on the size and nature of the company's business. However, for the remainder of this section, the above classification system will be used.

THE SYDNEY, AUSTRALIA, OPERA HOUSE

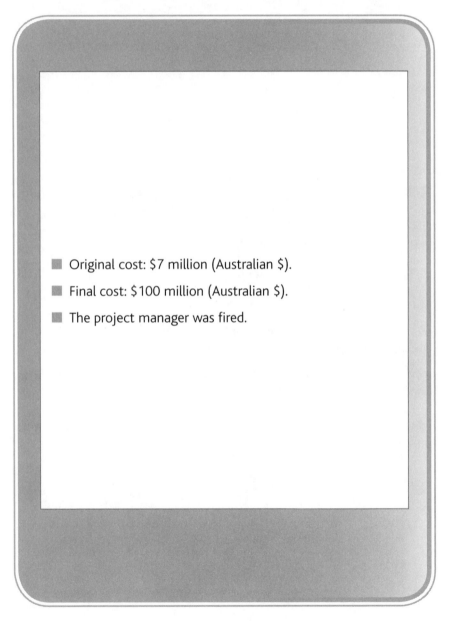

■ Original cost: $7 million (Australian $).

■ Final cost: $100 million (Australian $).

■ The project manager was fired.

The perceived value of a project can change over time. What was viewed initially as marginal value could end up a great value. The reverse is certainly true as well.

For example, the Opera House in Sydney, Australia, was a whopping 3,000 percent over budget, and the project manager was fired. Obviously, with such a large cost overrun, there were some stakeholders who argued that the originally perceived value of the project was lost, and that the project should have been canceled. Now, let's look at the next illustration and see how the perception of value had changed.

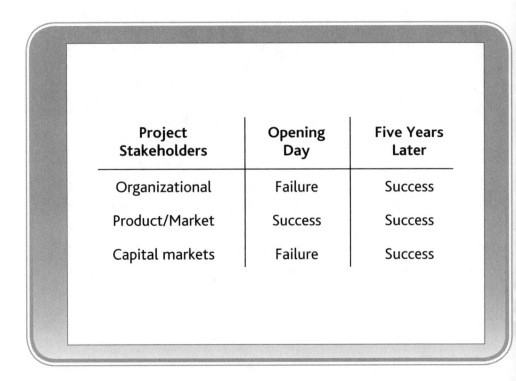

Project Stakeholders	Opening Day	Five Years Later
Organizational	Failure	Success
Product/Market	Success	Success
Capital markets	Failure	Success

In this illustration, we can see that today the project is considered to be extremely successful to the point where some people actually consider the Opera House the "Eighth Wonder of the World." The value is certainly there.

Cost overruns and scope changes could be considered acceptable, depending on the reasons and if there is a perception and a high probability that a greater future value may be achieved. Projects should be canceled when the key decision makers are convinced that the anticipated value, present and future, will not be attained. But sometimes, **the true value may not become evident until well into the future, and early cancellation of a project could be a mistake.** These types of decisions are difficult, but the key point here is to make sure that current and future value, as perceived by the customer and the supplier organization, are considered before making the decision. Some element of risk is involved, along with some forward-thinking management and leadership.

APPLE'S LISA COMPUTER

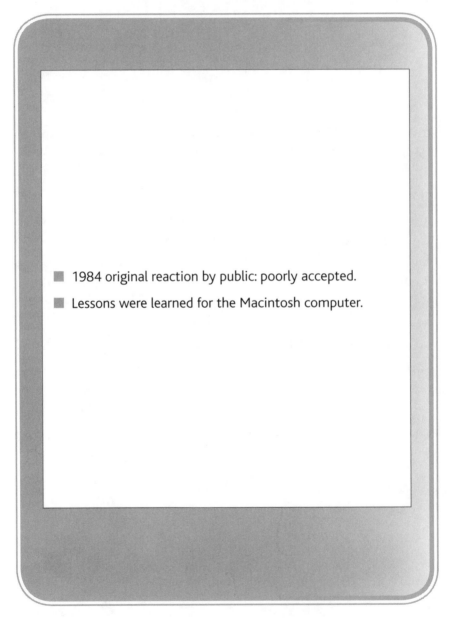

- 1984 original reaction by public: poorly accepted.
- Lessons were learned for the Macintosh computer.

In 1984, Apple launched the Lisa computer. Initially, the computer was poorly received. But was Lisa a success or a failure? The launch of the Lisa computer added credibility to Apple and enhanced their image as a computer manufacturer.

Value can also be found in a project that failed if lessons learned and best practices are discovered such that future projects will benefit. This was the case with the Lisa computer. Apple developed significant intellectual property in computer design and manufacturing, which enabled them to create a strong position in the marketplace. The value here was realized in terms of continuous improvement, flexibility, and innovation.

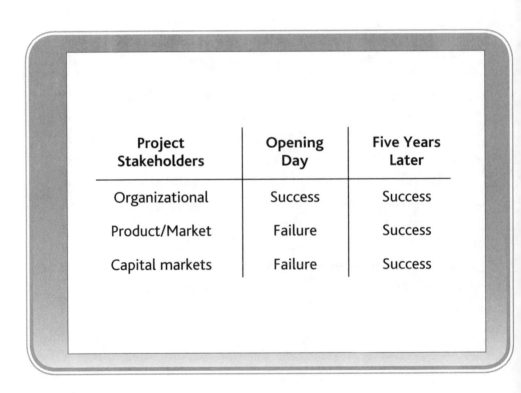

Project Stakeholders	Opening Day	Five Years Later
Organizational	Success	Success
Product/Market	Failure	Success
Capital markets	Failure	Success

As shown in this illustration, the launch of the Lisa computer was viewed initially by some as a failure. But five years later, Lisa was viewed as a success because the valuable lessons learned were applied to the launch of the Macintosh computer. It is possible that the Macintosh computer could have evolved by itself, but the Lisa computer certainly accelerated the learning process.

Some people argue that the only true project failures are those from which no knowledge is gained. **Failures are not really failures if intellectual property is gained that will benefit all future projects.** Therefore, a key component in the quest for value is the practice of reviewing, documenting, sharing, and acting on lessons learned and continuous enhancement of best practices.

DENVER INTERNATIONAL AIRPORT

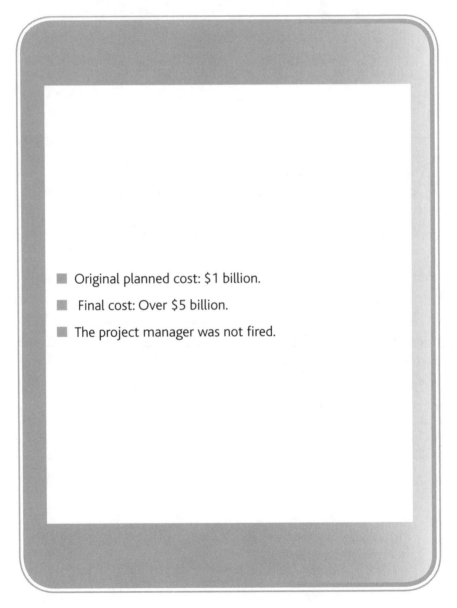

- Original planned cost: $1 billion.
- Final cost: Over $5 billion.
- The project manager was not fired.

D enver International Airport (DIA) was a project that, initially, the majority of stakeholders felt was not necessary. Only a few stakeholders believed that a new airport in was needed in Denver to handle the expected increase in passenger traffic. But that increase in traffic, even if it did occur, would probably happen gradually over 20 or more years.

DIA was eventually opened, almost two years late and 400 percent over budget. The problems all revolved around the complex baggage handling system that was designed for Concourse B (the United Airlines concourse) and then extended to encompass the entire airport. Some people argued that the airport was not worth the cost and that airport bankruptcy was possible if the airport could not service its debt load of about $1 million per day in interest payments alone. Others argued that the airport should have been opened as planned, but with a manual baggage handling system in place, so that revenue would be generated earlier to begin servicing the debt. The two-year delay increased the debt by more than $2 billion.

Project Stakeholders	Opening Day	Five Years Later
Organizational	Success	Success
Product/Market	Failure	Success
Capital markets	Failure	Success

The decision made by the City of Denver and the contractors was to keep the airport closed until the computerized baggage handling system was fully operational. This generated a large portion of the cost overrun but maximized the value of the airport well into the future. Today, travelers through the airport marvel at its success, and not many people still remember the cost overruns or the problems with the baggage handling system.

DIA provides us with an important lesson to consider: **Sometimes it is better to accept a cost overrun in order to maximize the value in a project's deliverables than to maintain the budget and add incremental value piecemeal over a decade or longer.** This is the same decision made by companies who must decide when to stop inventing, launch a product, and save the additional value for the next-generation product.

These three examples have shown that **managing projects for value** can lead to better results and significantly higher levels of stakeholder satisfaction than managing to the limitations of the triple constraint.

BALANCING STAKEHOLDERS' NEEDS

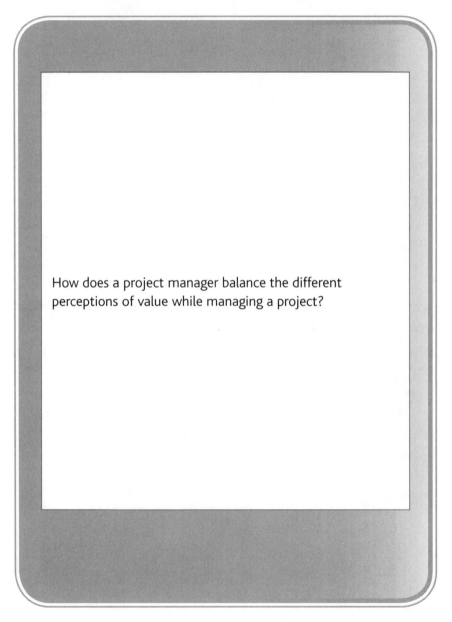

How does a project manager balance the different perceptions of value while managing a project?

Project managers must deal with a multitude of stakeholders, all with varying perceptions of value and at different levels of management internally. Balancing stakeholders' needs can be extremely difficult. Making a decision to maximize the value of one stakeholder could severely alienate other stakeholders. As an example, a project manager believes that he can exceed the customer's specification requirements and add significant value to the deliverable. This would also enhance the project manager's image in the company.

But to do this, the project manager would need to replace the "average" employees assigned to his project with the superior "subject matter experts," all of whom are currently assigned to other, higher-priority projects. This would benefit his project and him personally, but possibly at the expense of other projects. There is also no guarantee that the ultimate customer (i.e., one of the stakeholders) would agree with the additional costs incurred by using higher-salaried employees.

"I don't know the key to success,
but the key to failure is trying to please everybody."
BILL COSBY - AMERICAN COMEDIAN, ACTOR, AUTHOR,
TELEVISION PRODUCER, AND ACTIVIST

As suggested by this quote, the project manager will occasionally be required to make decisions that benefit some stakeholders more than others, and there is a risk that some decisions will result in conflict. The project manager and team should ensure that all stakeholders are identified and that the influence of each stakeholder as it relates to the project is fully understood.

TRADITIONAL CONFLICTS OVER VALUES

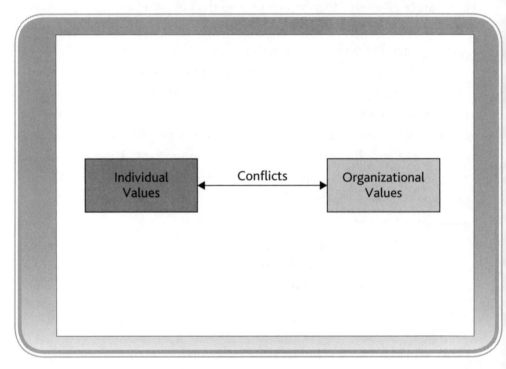

Sooner or later, everyone is placed in a position where he or she must decide what is more important—what they perceive as value gained personally or what the organization will gain in terms of value. This situation forces people to make very difficult decisions. A quote from the movie *Star Trek II: The Wrath of Khan* comes to mind here:

"The needs of the many outweigh the
needs of the few."

MR. SPOCK - FICTIONAL CHARACTER IN THE STAR TREK
TELEVISION AND MOVIE SERIES

This particular quote relates well to the situations often faced by project managers who must occasionally make decisions that will impact some stakeholders very negatively.

These types of conflicts can permeate all levels of management. Executives may make decisions in the best interest of their pension rather than the best interest of their firm. One executive wanted to be remembered in history books as the pioneer of high-speed rapid transit. He came close to bankrupting his company in the process of achieving his personal perception of value at the expense of the corporation.

Another example is the project manager who was quite disheartened when management canceled his research and development (R&D) project because they could not recognize the long-term value in what he was doing. The project manager believed that the value of this project would be in the best interest of the company if completed. He kept working on this project using charge numbers from other projects. When management discovered this, the project manager was fired even though he believed he was making decisions in the best interest of the company.

PROJECT MANAGEMENT VALUE CONFLICTS

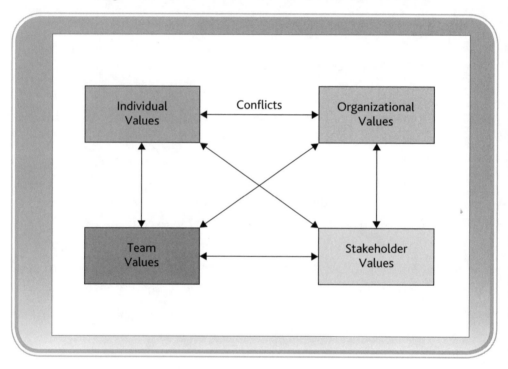

In the previous illustration, we discussed the conflict between individual and organizational values. In a project environment, this conflict is more complex, as seen in the following illustration. There are now competing conflicts between:

- Individual values

- Organizational values

- Team values

- Stakeholder values

The project manager must work with all of these groups and balance, if possible, each group's perception of value. The next illustration shows how different perceptions can occur.

VALUE PERCEPTIONS WITHIN A PROJECT

Project Manager	Team	Organization	Stakeholders
<ul><li>Accomplishment of Objectives</li><li>Creativity</li><li>Innovation</li></ul>	<ul><li>Achievement</li><li>Advancement</li><li>Ambition</li><li>Credentials</li><li>Recognition</li></ul>	<ul><li>Continuous Improvements</li><li>Learning</li><li>Quality</li><li>Strategic Focus</li><li>Morality and Ethics</li><li>Profitability</li><li>Recognition and Image</li></ul>	<ul><li>Organizational: Job Security</li><li>Product/Market: Quality and Performance</li><li>Financial: Financial Growth</li><li>Product Usefulness</li></ul>

This illustration shows the value perceptions within a project. What is interesting about this illustration is that project managers and team members focus heavily on internal or personal values. When project managers and team member are assigned to a new project, they often ask: "What's in it for me if I accept this assignment?" Basically, they are trying to identify what value they will receive personally from this assignment. This is a typical situation encountered by project managers. The issue here is to show that personal value can be attained by actually accepting and committing to the project objectives and the higher-level strategic organizational objectives and goals.

Chapter

THE
COMPONENTS
OF SUCCESS

FOUR CORNERSTONES OF SUCCESS

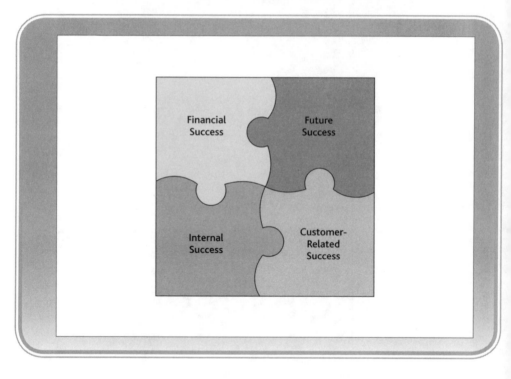

Defining project success has never been an easy task. The focus has, for most projects of the past, always been the triple constraint—balancing cost, scope, and schedule and meeting the set objectives. Today, we believe that there are four cornerstones for success:

- **Internal success:** The ability to have a continuous stream of successfully managed projects using an enterprise project management methodology and that continuous improvement occurs on a regular basis

- **Financial success:** The ability to create a long-term revenue stream that satisfied the financial needs of the key stakeholders

- **Future success:** The ability to produce a stream of deliverables that will support the future existence of the firm

- **Customer-related success:** The ability to satisfy the needs of the customers over and over again to the point where you receive repeat business and the customers treat you as though you are a partner rather than a contractor or supplier

CATEGORIES OF SUCCESS

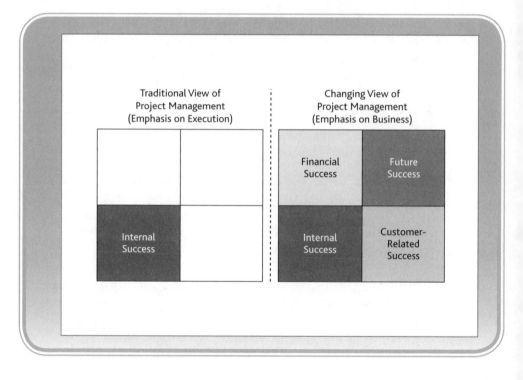

Historically, the triple constraint focused on only internal success. But, as we showed, meeting the imposed conditions of the triple constraint alone does not guarantee success.

Some projects can support more than one quadrant, as shown in the illustration on the previous page. For example, developing a product using a company's proprietary technology and having the product well accepted in the marketplace can lead to financial, future, and customer-related success. Another example would be the creation of an enterprise project management (EPM) methodology. Although the value of this can be regarded as internal success, the methodology can lead to a long-term relationship with customers and eventually financial success. Therefore, the success and ultimate value achieved can impact more than one quadrant.

CATEGORIES OF VALUES

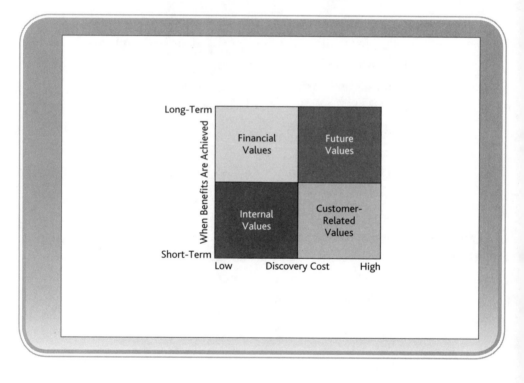

Since the two primary components of success are the triple constraint and the value produced, we can argue that success can be better defined as value achieved. Considering that the triple constraint can be applicable to all of the quadrants for the remainder of this module, we will focus on values achieved.

The illustration on the previous page also shows the relationship between the benefits achieved from the value as opposed to the investment or discovery cost to achieve the values. As an example, in the short term, there is a low discovery cost to improve the company's processes (i.e., internal values), but a potentially high cost to maintain customer satisfaction. Long-term values support financial success and future success.

Achieving value can always be accomplished but perhaps at an unusually high cost. Executives must carefully compare the cost of obtaining the desired value with the value that is actually realized (at project completion and future potential value). Spending $2 million to achieve a deliverable that provides $1 million in value is not a good business decision. Of course, placing a dollar figure to value can be somewhat difficult, if not impossible, to do—at least with reasonable accuracy.

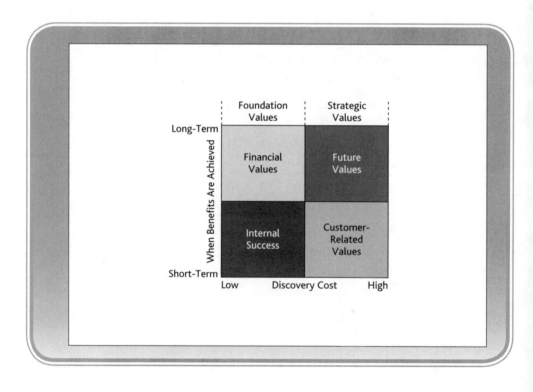

Previously, we identified two major categories of values; foundation values and strategic values. The foundation values can be divided into two parts—internal values and financial values—whereas the strategic values are more aligned with future values and customer-related values.

- **Foundation values** are those values that must be achieved in the short term for the continuous operation of the firm on a day-to-day basis. This includes methodologies and processes to support ongoing activities. Cash flow is also needed to continue operations, so some activities that provide financial value are needed.

- **Strategic or innovation values** are those values that must be achieved for the long-term survivability of the firm. This includes maintaining a strong list of clients, especially those that treat you as a potential partner, and having a pipeline of new projects that support the future products and services of the firm.

DECIDING ON THE QUADRANT

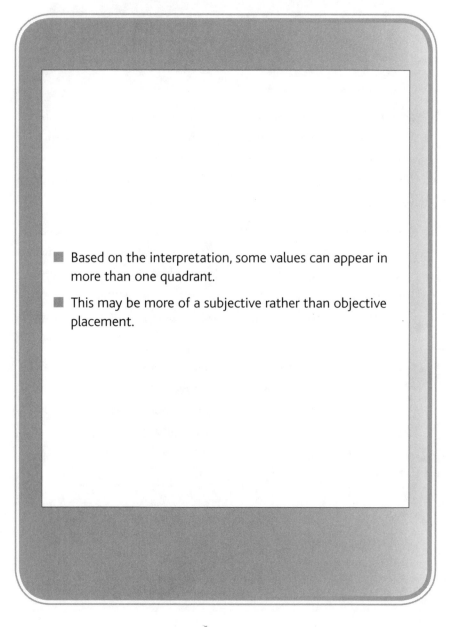

- Based on the interpretation, some values can appear in more than one quadrant.
- This may be more of a subjective rather than objective placement.

As can be seen from the preceding illustration, deciding in which quadrant to place a given project is very subjective. Some projects can encompass more than one quadrant and possibly even all four quadrants.

Because of this subjectivity, we will place certain projects or activities in the quadrant where maximum value will be or can be achieved. For example, consider a company that successfully improves its process for commercialization of a product. We could argue that it is an internal value because of process improvement; a financial value because it should allow us to increase our margins; a customer-related value because we can get the needed products to the customers more quickly; or a future value because it is designed to improve potentially all products or services in the future. In our minds, this would fall under internal value.

INTERNAL VALUES

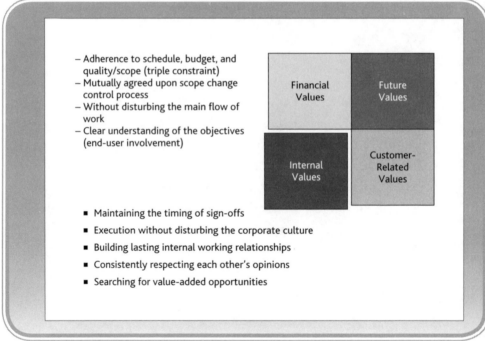

- Adherence to schedule, budget, and quality/scope (triple constraint)
- Mutually agreed upon scope change control process
- Without disturbing the main flow of work
- Clear understanding of the objectives (end-user involvement)

Financial Values

Future Values

Internal Values

Customer-Related Values

- Maintaining the timing of sign-offs
- Execution without disturbing the corporate culture
- Building lasting internal working relationships
- Consistently respecting each other's opinions
- Searching for value-added opportunities

Internal values focus heavily on the development of an enterprise project management (EPM) methodology and using it effectively without disturbing the organization's ongoing work. The methodology undergoes continuous improvement in order to improve the efficiency and effectiveness of the organization.

Although the triple constraint can be argued as belonging to all four quadrants, it is subjectively included here under internal values. Internal values include corporate culture support such as teamwork, communication, cooperation, and trust.

FINANCIAL VALUES

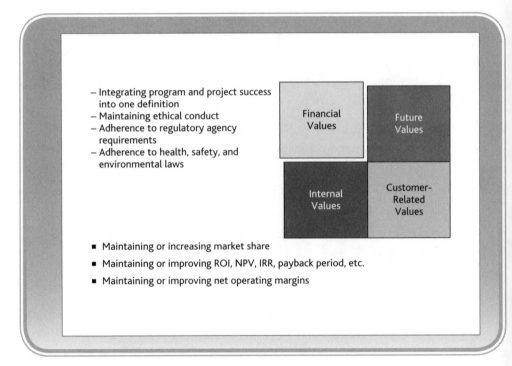

– Integrating program and project success into one definition
– Maintaining ethical conduct
– Adherence to regulatory agency requirements
– Adherence to health, safety, and environmental laws

Financial Values

Future Values

Internal Values

Customer-Related Values

- Maintaining or increasing market share
- Maintaining or improving ROI, NPV, IRR, payback period, etc.
- Maintaining or improving net operating margins

Financial values focus on the long-term financial health of the company. Adhering to government regulations concerning health, safety, environmental issues, and ethical behavior also effects long-term financial health by avoiding penalties, lawsuits, tarnishing the corporate image, and losing customers.

One or two successful projects do not guarantee long-term financial health but may be needed in the short term to support cash flow requirements. Financial health can be defined as a stream of successful projects that meets the company's financial targets. The proper selection of projects is essential to achieve and maintain financial health thus implying successful implementation of a project management office (PMO) and a project portfolio management process.

FUTURE VALUES

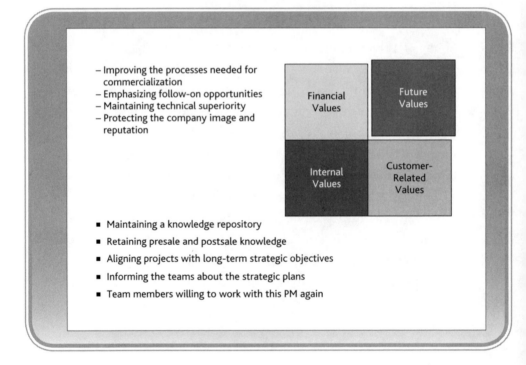

- Improving the processes needed for commercialization
- Emphasizing follow-on opportunities
- Maintaining technical superiority
- Protecting the company image and reputation

Financial Values

Future Values

Internal Values

Customer-Related Values

- Maintaining a knowledge repository
- Retaining presale and postsale knowledge
- Aligning projects with long-term strategic objectives
- Informing the teams about the strategic plans
- Team members willing to work with this PM again

Future values are also heavily aligned with project selection and the portfolio management of projects. While some project selection processes focus only on near-term success, some projects must be selected for their alignment with long-term strategic goals and where profitability and value may not be recognized until some time in the future. Financial models such as decision trees, sensitivity analysis, and internal rate of return (IRR) could be used to assess future values. It is important to consider the impact of the global economic situation and the emergence of companies in third world countries as potential competitors.

CUSTOMER-RELATED VALUES

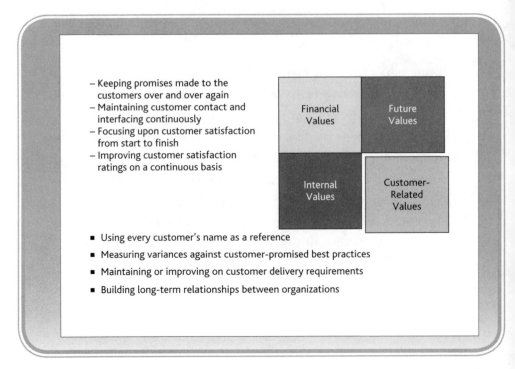

– Keeping promises made to the customers over and over again
– Maintaining customer contact and interfacing continuously
– Focusing upon customer satisfaction from start to finish
– Improving customer satisfaction ratings on a continuous basis

Financial Values

Future Values

Internal Values

Customer-Related Values

- Using every customer's name as a reference
- Measuring variances against customer-promised best practices
- Maintaining or improving on customer delivery requirements
- Building long-term relationships between organizations

Because of the subjectivity involved, many of the above customer-related values could have been placed in other quadrants. However, for simplicity's sake, customer-related values allow the company to:

- Provide additional products/service to existing customers

- Develop new products based on customer input

- Develop long-term partner relationships with existing clients and create new partnerships with other clients

- Maintain a well-respected corporate image and brand recognition that attracts new customers

- Focus on creating and maintaining a leadership position

- Maintain high levels of customer satisfaction

REASONS FOR INTERNAL VALUE FAILURE

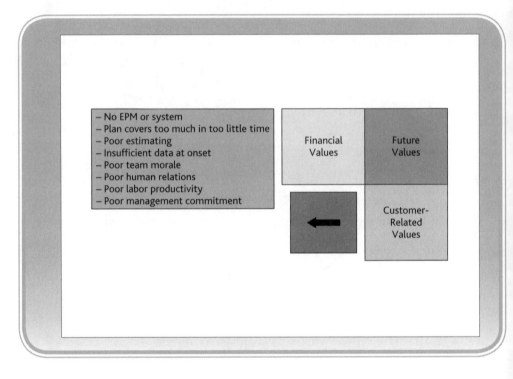

The failure to achieve internal value may be attributed to a project management methodology that isn't working correctly. For years, blame was placed on quantitative factors such as poor planning, estimating, scheduling, and cost control. There were also valid arguments about failure relating to situations where there was no EPM methodology in place.

Today, more than ever, we are recognizing that failure can also result from behavioral issues, as seen in the preceding illustration. Organizations in which project management thrives on teamwork, effective communications, cooperation, and trust can overcome most of the quantitative failure issues. It is important to note that without positively influenced human behavior during project planning and execution, the effectiveness of the resources will be diminished and even the best plans can fail.

REASONS FOR FINANCIAL VALUE FAILURE

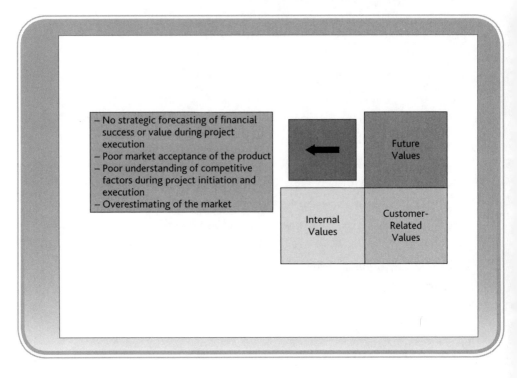

Long-term financial failure can be devastating to the health of a company. Other than the most common reasons, which appear in the illustration on the previous page, we can also include:

- Poor strategic planning

- Poor portfolio management techniques

- Poor identification and tracking of assumptions and constraints

- Poor understanding on the market

- Overly inflated financial expectations

Even though a company may possess an outstanding enterprise project management methodology, these failure factors can still exist. Previously, we stated that if the wrong project is selected, it is irrelevant as to how well or how poorly the projects are managed.

REASONS FOR FUTURE VALUE FAILURE

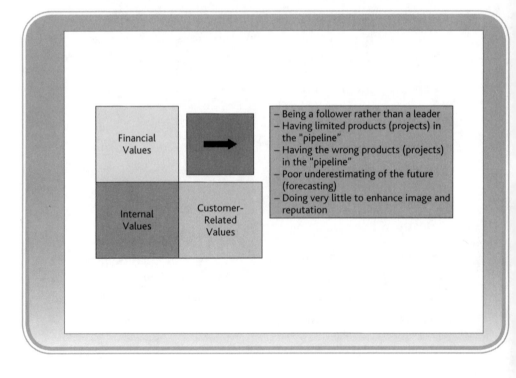

Future value failure can be caused by the same items shown and discussed in the illustration on the previous page. But there are additional issues, the five most common being:

- The selection of projects that fail to enhance the firm's image, reputation, or credibility

- The desire to be a marketplace follower or copycat rather than a market leader

- Failing to invest in the proper research and development (R&D) activities

- Failing to maintain a policy of continuous improvement

- Failing to listen to the "voice of the customer" when developing new products and services

REASONS FOR CUSTOMER-RELATED VALUE FAILURE

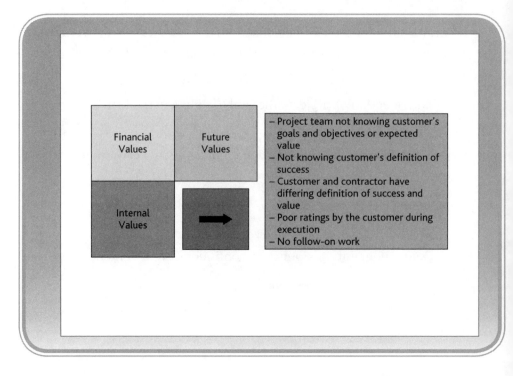

Customer-related value failures can emanate from poor customer communications. Examples can include:

- Not understanding the customer's requirements

- Not providing the customer with meaningful status reporting during project execution

- Not conducting phase-end reviews and continued scope verification

- Not obtaining feedback from the customer during execution and after project closure

- Not making the customer feel important (including the customer in planning and decision making)

- Not establishing and maintaining an agreed-upon change control process

Other issues appear in the preceding illustration.

ANTARES SOLUTIONS

Financial Values	Future Values
▪ Using the EPM system throughout the enterprise, not just in the Information Technology Group	▪ Enterprise-wide PM training ▪ Periodic project audits to validate the EPM and seek improvements ▪ Closure presentations
Internal Values	**Customer Values**
▪ Consistent structure for each project (EPM) ▪ Using online tools for communicating plans ▪ Maintaining an official project issues list	▪ Using a formalized change control process that includes the executive steering committee

The following several illustrations show examples of perceived value for a specific company or industry.* Antares Management Solutions** is representative of information technology (IT) organizations. Management decided to embark upon significant changes in the way they manage projects. A document was prepared and fully supported by senior management, illustrating the expected values and copies were provided to all employees. The values are shown in the preceding illustration.

What is interesting about the entries in each quadrant is that the definitions of success, values, and best practices need not be complex. The entries in the quadrants reflect more on the things the company must do well to achieve these values.

---

*The examples provided on the next several pages are based upon the authors' interpretation of the limited information provided in previously published texts discussing best practices in project management. Actual positioning of information in each of the quadrants can change based upon economic conditions.
**Source: Harold Kerzner, *Advanced Project Management: Best Practices on Implementation*, 2nd ed. Hoboken, NJ: John Wiley & Sons, © 2004, pp. 136–138.

GENERAL ELECTRIC (PLASTICS GROUP)

Financial Values	Future Values
▪ Adhering to environmental regulations ▪ Adhering to safety regulations	▪ Improvements in technology ▪ Delivering productivity for manufacturing operations
Internal Values	**Customer Values**
▪ Time ▪ Cost ▪ Quality	▪ Customer satisfaction

General Electrics Plastic Group* focuses heavily on capital improvement projects, but some projects were for external clients. General Electric has eight components as part of their success criteria, as shown in the preceding illustration.

---

*Source: Harold Kerzner, *Advanced Project Management: Best Practices on Implementation*, 2nd ed. Hoboken, NJ: John Wiley & Sons, © 2004, p. 32.

ASEA BROWN BOVERI (ABB)

Financial Values	Future Values
■ Using the EPM system, achieving booked or better profit margins	■ The PMO conducts periodic audits of each project to validate the EPM and seek improvements
Internal Values	**Customer Values**
■ Predictable project management (EPM) ■ Delivery, delivery, delivery—on time delivery within budget —every time	■ Complete customer satisfaction

A sea Brown Boveri (ABB)* is a multinational corporation in a variety of businesses, including heavy construction. The preceding
illustration shows the drivers for their values.

---

*Source: Harold Kerzner, *Advanced Project Management: Best Practices on Implementa-
tion*, 2nd ed. Hoboken, NJ: John Wiley & Sons, © 2004, pp. 33, 44–45.

WESTFIELD GROUP

Financial Values	Future Values
	■ Need to provide complete business solutions rather than just products
Internal Values	**Customer Values**
■ Consistent structure for each project (EPM) ■ Using the online intranet EPM system	■ Recognizing that the needs of the customers had changed ■ Recognizing changing customer relationships ■ Customer partnering

Westfield Insurance*, a Division of Westfield Group, recognized that project management growth was essential for corporate success. According to a spokesperson,

> "Westfield Insurance recognized that the needs of our customers had changed with regard to information technology and information systems projects. Also, the relationship with our customers had changed. Our customers were now expecting complete solutions to their business needs rather than just products. We needed an organization that was designed to partner with our business customers and specialized in providing high-quality business solutions."

Since the majority of the customers were internal to the company, profit margins were not assigned to the IT projects, which is why the financial value box is empty.

---

*Source: Harold Kerzner, *Advanced Project Management: Best Practices on Implementation*, 2nd ed. Hoboken, NJ: John Wiley & Sons, © 2004, pp. 138–142.

COMPUTER ASSOCIATES TECHNOLOGY SERVICES

Financial Values	Future Values
■ Maintaining net operating margins ■ Assigning qualified resources	■ Management of customer expectations ■ Capturing best practices ■ Measuring variances against best practices
Internal Values	**Customer Values**
■ Proper scoping of the customer's requirements ■ Handling exceptions ■ Weekly measurement of progress versus the baseline	■ Customer satisfaction ratings ■ Maintaining customer communications and involvement ■ Sharing risk management information

Computer Associates (CA)* is a multinational IT organization. They maintain a reasonably large project management office and are strong believers in capturing best practices in project management. Prior to 2005, their best practices library consisted of 177 best practices, which were both process- and product-related best practices. After 2005, the number of best practices was scaled back to 151. The reason for this was either:

- The best practice is no longer a best practice and has been retired.

- The best practice has been combined with other best practices.

---

*Source: Harold Kerzner, *Advanced Project Management: Best Practices on Implementation*, 2nd ed. Hoboken, NJ: John Wiley & Sons, © 2004, pp. 68–72, 535–549.

CONVERGENT COMPUTING

Financial Values	Future Values
■ Having experienced and well-rounded technical resources	■ Having well-defined policies and processes that leverage best practices ■ Having formalized EPM processes
Internal Values	**Customer Values**
■ Proper scoping of the customer's requirements ■ Having carefully crafted teams with well-defined roles and responsibilities	■ Leveraging our experience and knowledge to achieve the deliverables ■ Enhanced colloboration and communication

Convergent Computing (CCO)* is a medium-sized, privately held company that partners with its clients to provide professional and knowledgeable IT consulting services. The road to success was difficult but finally achieved. According to a spokesperson from CCO,

> "Many of the consultants and engineers didn't understand why they should have to learn and or deal with the project management and this made it difficult to get them 'on board.' Many of them did not want to take the additional time to learn project management methodologies and felt they had enough on their hands just staying knowledgeable with the ever-changing technology. When project managers were assigned to projects and began scheduling kick-off meetings or asking for a regular status update, consultants felt they were burdened with additional tasks or asks to attend meetings they didn't see as necessary."

---

*Source: Harold Kerzner, *Project Management Best Practices: Achieving Global Excellence*, 1st ed. Hoboken, NJ: John Wiley & Sons, © 2006, pp. 43–44.

MOTOROLA

Financial Values	Future Values
■ Control to prevent costly scope creep ■ Having the right skills available when needed ■ Unequaled technical execution	■ Maintaining high-quality standards
Internal Values	**Customer Values**
■ Well-defined processes with formalized gate reviews ■ Commitment to schedules	■ Early customer involvement ■ Effective customer communications

M otorola* is a company that consistently strives to improve its
process. According to a spokesperson,

"What worked well last year or this year, may not meet our
needs in the future. We have goals to improve the effectiveness
of project performance and product development. We believe
that the implementation of the project management process is a
key to achieving our goal [and value]. We intend to be the best
in class in project management and are investing resources to
this end."

---

*Source: Harold Kerzner, *Advanced Project Management: Best Practices on Implementation*, 2nd ed. Hoboken, NJ: John Wiley & Sons. © 2004, pp. 571–573.

AUTOMOTIVE SUPPLIERS SECTOR

Financial Values	Future Values
<ul><li>Maintaining net operating margins</li><li>Development of a project management cost model for each discipline based on industry benchmarks</li></ul>	<ul><li>Full business review processes to ensure lean disciplines</li><li>Lean supply chain management</li><li>Use an opportunity tracker system</li></ul>
Internal Values	**Customer Values**
<ul><li>Global common EPM applications including QS9000 and Six Sigma</li><li>EPM alignment to business objectives</li><li>World-class life-cycle management (EPM)</li></ul>	<ul><li>Prevent quality spills to the customer</li><li>A structured change management process</li><li>On-site customer liaisons</li></ul>

The preceding illustration represents the automotive suppliers sector. With limited customers for the automotive suppliers and a large number of suppliers, profit margins are squeezed, and the primary customers (i.e., Ford, Chrysler, and General Motors) want higher quality at a lower price.

THE INDUSTRY AS A WHOLE IS PRESSURED BY:

- Customers wanting lower prices

- Customers wanting higher levels of quality

- Customers demanding the development of a world-class project management methodology

- Customers mandating that the suppliers adhere to rigid quality standards

BANKING SECTOR

Financial Values	Future Values
■ Focusing on core businesses ■ Managing credit risk ■ A standard EPM system to provide the best ROI and cancel if necessary	■ Gathering and sharing lessons learned ■ Correct targeting of acquisitions and growth investments
Internal Values	**Customer Values**
■ Consolidates the technologies and processes that overlap ■ Consistent EPM system across all lines of business	■ Include customers in the stage-gate process ■ Providing accessible customer-centric banking options ■ Cross-sell products

The preceding illustration is an example of the banking sector. The sector is made up of many banks, many customers, and numerous mergers and acquisitions. Note that each set of values is actually part of a value system that addresses the enterprise needs.

The banking sector is sensitive to risk, and the values associated with financial institutions are related to the desire to create a safe environment for their clients. However, recent economic events have changed the core businesses of many of the banks, and these changes can significantly alter the elements in each quadrant.

COMMODITY PRODUCTS (MANUFACTURING) SECTOR

Financial Values	Future Values
■ Compliance with Sarbanes-Oxley laws ■ Compliance with local, state, and federal laws ■ Having a common financial platform	■ Technical innovation capability ■ Using project management to obtain a sustained competitive advantage
Internal Values	**Customer Values**
■ Designing new facilities for health, safety, and the environment ■ An employee recognition system	■ New business awards ■ Supplier awards ■ Having a customer-focused quality assurance process ■ Customer satisfaction

The preceding illustration represents a typical commodity manufacturing firm. In most cases, compliance to federal and state laws takes precedence in providing value. The sector is also in favor of improving brand name and corporate image by receiving awards. Some companies advertise and focus on the marketability of their awards, such as quality awards, performance awards, and customer satisfaction awards. This is why companies often spend millions of dollars to become ISO 9000 certified, achieve Six Sigma levels of performance, or to win the Malcolm Baldrige or other prestigious awards.

LARGE COMPANIES

Financial Values	Future Values
<ul><li>Multiple product lines</li><li>Paying out dividends</li><li>Project that become "cash cows" and "stars"</li></ul>	<ul><li>Successful innovation</li><li>Being a market leader</li><li>Product development (innovation, patents, and trade secrets)</li><li>Image/reputation</li></ul>
Internal Values	**Customer Values**
<ul><li>Product development process</li><li>A structured project review process</li><li>Identifying "exit" champions</li></ul>	<ul><li>Customer satisfaction ratings</li><li>Maintaining customer relations</li><li>Customer partnerships</li></ul>

Large companies, especially those that are cash rich, can invest heavily in R&D, product enhancements, and process improvements. These companies can accept a financial loss and project failures in hopes of developing better customer relations in the future.

Many large companies believe that the key to the future lies in the company's intellectual property, which can enhance the company's image. Cash-rich companies may be in a position to accept a 50 percent success rate for their new products, whereas cash-poor companies may accept nothing less than 80 percent or more, and are more selective in the projects they accept.

SMALL COMPANIES

Financial Values	Future Values
■ Maintain market share growth for survival ■ Short-term profit-driven strategy ■ Affordable cost of quality and R&D	■ Maintaining independence ■ Having a single goal that supports a niche strategy ■ Selective bidding
Internal Values	**Customer Values**
■ An EPM system that focuses more on enhancements rather than new product development ■ Well-structured risk management process	■ Customer satisfaction ratings ■ Customer retention

Smaller companies perceive value differently than large companies. To understand why, consider the following criteria that small companies often use as part of strategic planning:

- Cash limitations force the company to be more selective in the projects they work on

- Project selection focuses on a niche strategy

- The risks of failure are given more attention than in a large company.

- Most customers are located regionally

- Customers retention is a high priority

- Growth comes from cash flow rather than borrowing

- The company may not be able to be the market leader and may be content in a position as a follower or copycat product company

Chapter

6

SUCCESS AND BEST PRACTICES

FROM VALUES TO BEST PRACTICES

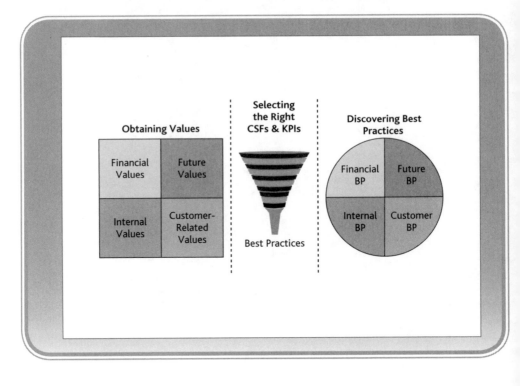

The ultimate goal of value-based project management is obviously to obtain the greatest value of the deliverable from the customer and the producing organization's perspective. In the process of doing so, best practices (BPs) can be discovered that allow the organization to create additional value in terms of efficiency, customer satisfaction, and enhanced products and services, especially when implementing future projects.

To do this, it is important to select the right critical successful factors (CSFs) and key performance indicators (KPIs). The definitions of these terms appear on the next two pages.

TWO COMPONENTS OF SUCCESS

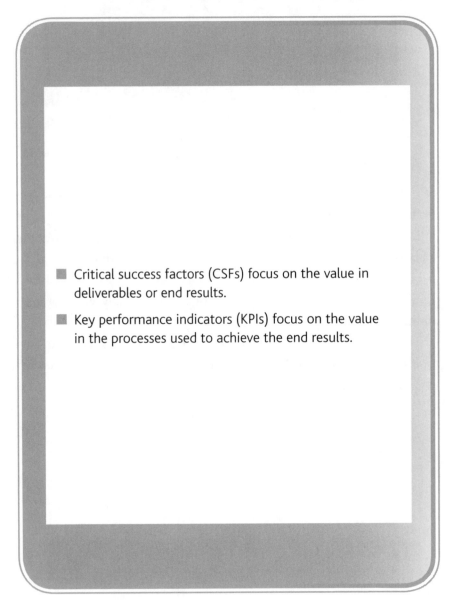

■ Critical success factors (CSFs) focus on the value in deliverables or end results.

■ Key performance indicators (KPIs) focus on the value in the processes used to achieve the end results.

CSFs and KPIs can be viewed as best practices if the right CSF and KPI are selected and applied. CSFs focus on the end result and its perceived value. KPIs focus on the processes to achieve the CSF.

CSFs are difficult to track. Usually, the CSF is identified at the end of the project. KPIs are tracked throughout the project. KPIs are the measurement instruments that provide some indicator that the CSF will be obtained.

A KPI is a financial or nonfinancial measure used to help an organization measure progress toward a stated organizational goal or objective.

A CSF would be something that needs to be in place to achieve that objective, for example, the launch of a new product or service.

KPI = Number of new customers

CSF = Installation of a call center for managing the customers

- CSFs are elements that are vital for a given strategy to be successful

- KPIs are measures that quantify objectives and enable the measurement of strategic performance

REDEFINING VALUE METRICS (CSFs AND KPIs)

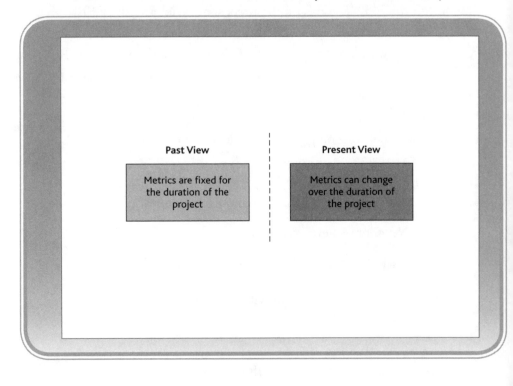

Generally, CSF and KPI metrics are established by the company and applied to all projects. This is particularly true when the definition of success is exclusively associated with the triple constraint. An example of a traditional CSF is adherence to schedules, budgets, quality, timing of sign-offs, and adherence to the change control process. Typical KPIs are use of the methodology and comparing progress against the baseline. These metrics may change, but the change will be slow.

When the definition of success also includes value, then value (CSF and KPI) metrics must also be defined. The value metrics can change over the duration of the project, especially if the perceived value changes.

THE NEED FOR CHANGING METRICS

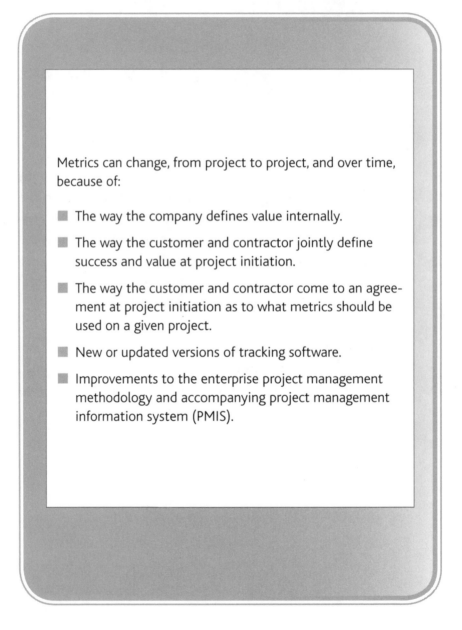

Metrics can change, from project to project, and over time, because of:

- The way the company defines value internally.
- The way the customer and contractor jointly define success and value at project initiation.
- The way the customer and contractor come to an agreement at project initiation as to what metrics should be used on a given project.
- New or updated versions of tracking software.
- Improvements to the enterprise project management methodology and accompanying project management information system (PMIS).

The preceding illustration shows the need to monitor and change metrics as the business environment changes. With traditional project management, the enterprise project management (EPM) system usually contains a standard set of CSFs and KPIs that are used on each and every project. It has been a common practice to establish the CSFs and KPIs around the triple constraint.

When value becomes part of the success criteria, and when there are multiple forms of value, the CSF and KPI can change from project to project. Changes can also occur as a result of changes to the EPM and project management information system (PMIS). Methodologies today must continue to adapt to changing values and their impact on CSF and KPI.

PROJECT MANAGEMENT OFFICE INVOLVEMENT

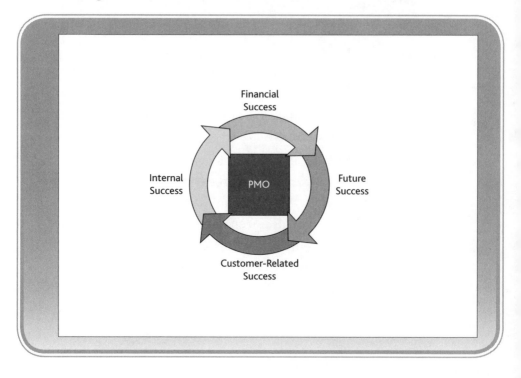

The selection of which set of metrics to use is not left to chance nor established by solely by the project manager. Most companies use the project management office to establish the metrics that will be used to asses project performance. The value metrics may be established through a team effort including the project management office, the project manager, and the customer.

Without an established project management office (PMO), life in the project management environment may become difficult. Metrics may not be established, lessons learned may not be identified, and BPs will not be captured. If mistakes were made, they may be repeated on future projects.

DISCOVERY OF BEST PRACTICES

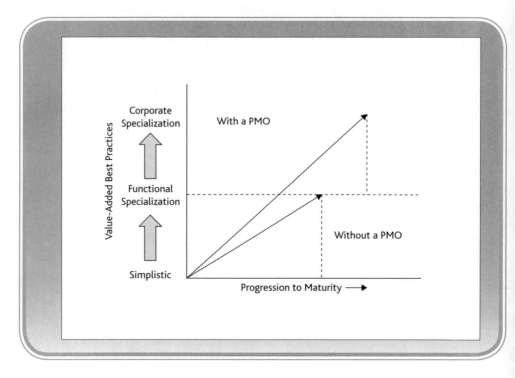

The importance of a PMO cannot be overlooked or ignored. This is shown in the preceding illustration. Without a PMO, the identification of BPs becomes a functional responsibility and there is a risk that BPs may not be captured, documented, or communicated effectively through an organization. The majority of discovered BPs are intended to benefit the entire organization and this may be difficult for a single-line organization to manage effectively.

With a PMO in place and staffed, it is easier to discover the value BPs. Also, with a PMO, additional BPs that could accelerate the project management maturity process may be discovered through project reviews, customer satisfaction reports, and informal discussions.

THE DEBRIEFING PYRAMID

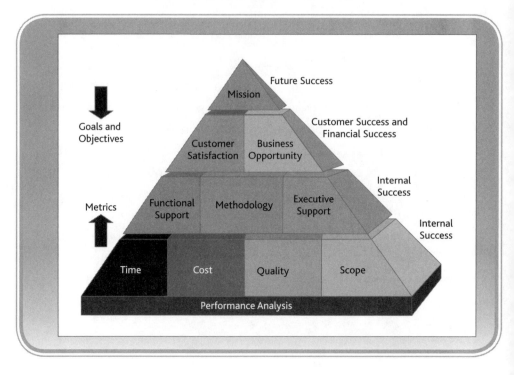

The preceding illustration identifies the debriefing pyramid. The PMO, the project team, and the functional groups work together to determine internal success factors. The PMO will work with marketing, sales, and senior management to determine customer, future, and financial success. This information is documented and utilized to improve existing processes and support continuous improvement.

DISCLOSURE OF BEST PRACTICES

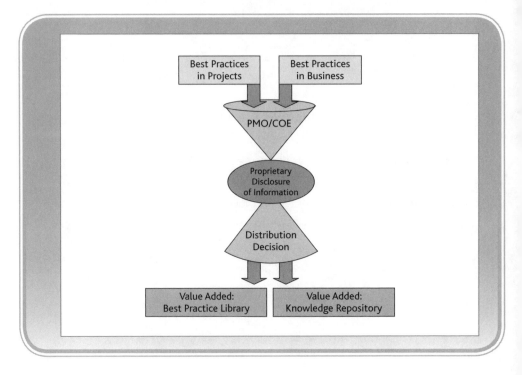

Previously, we stated that today's project managers are expected to capture BPs in business as well as BPs in the management of projects. This can lead to complications if the company maintains a BP library that contains proprietary information. In this case, it is important to differentiate BPs in the library that may be shared with customers and those that are intended for use by organizational employees.

To resolve this problem, companies have begun classifying documented BPs as propriety knowledge or nonpropriety knowledge. Value-added BPs that will be shared with the clients are placed in a "common" or BP library. The proprietary BPs are placed into a knowledge repository, which is intended for internal use only.

LEVELS OF SUCCESS IN OBTAINING VALUES

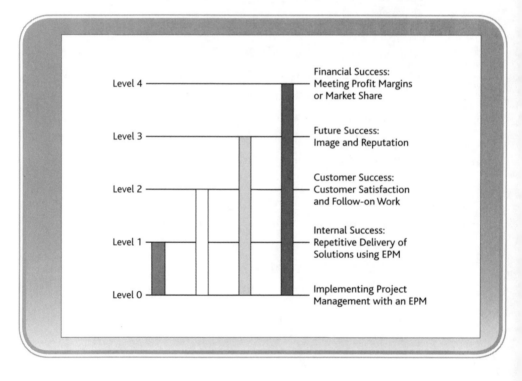

We showed earlier that there were benefits or value that could be obtained in the short term when a project is completed, whereas other values may take years before they are realized. The preceding illustration shows the typical order or level for the values. Level is the implementation of project management using an enterprise project management methodology. Levels 1 and 2 are near-term levels, whereas Levels 3 and 4 are long term.

PROJECT MANAGEMENT KNOWLEDGE

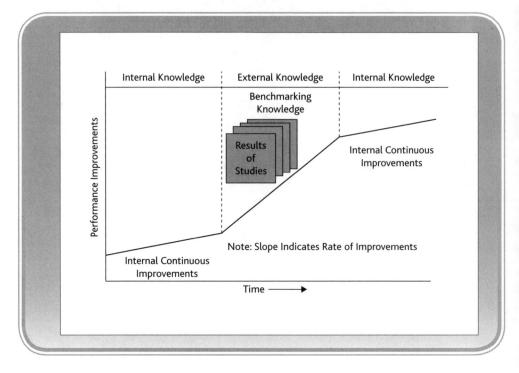

The purpose behind capturing and sharing practices is to improve organizational performance. Continuous improvement is essential in the business and project environment and should become part of the organization's culture (Toyota, as an example, has a definite culture of continuous improvement that is evident in every level of management and business unit). Sometimes bench marking can provide an organization with sufficient knowledge to develop additional best practices that will enable continuous improvement efforts to occur more quickly than using only internally captured BPs.

Companies that always promote from within the company create an inbreeding culture to the point where new ideas may not be forthcoming. Benchmarking, literature studies, and industry analyses can generate excellent ideas. **However, it should be understood that a BP in one company may not be an applicable BP in other companies. Not all BPs are transferable.**

PROJECT MANAGEMENT BENCHMARKING

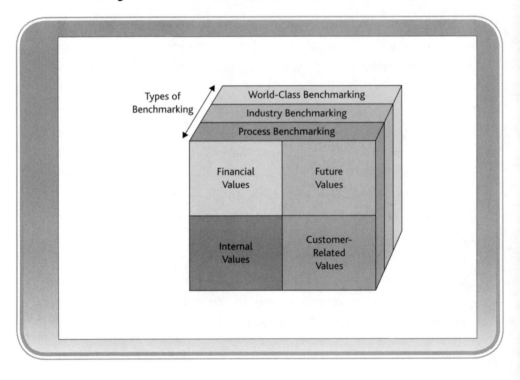

If value-oriented benchmarking is to be performed, it can be done according to these three types shown in the illustration on the previous page. In general, each type of benchmarking may produce different results, such as:

- Process benchmarking: Discovery of internal values

- Industry benchmarking: Discovery of internal and customer-related values

- World-class benchmarking: Discovery of internal, customer-related, future, and financial values

The cost of performing world-class benchmarking is significantly greater than performing process or industry benchmarking. For significant results to be obtained, the company must be willing to commit time, effort, and financial resources to the process.

SHARING VALUES DURING BENCHMARKING

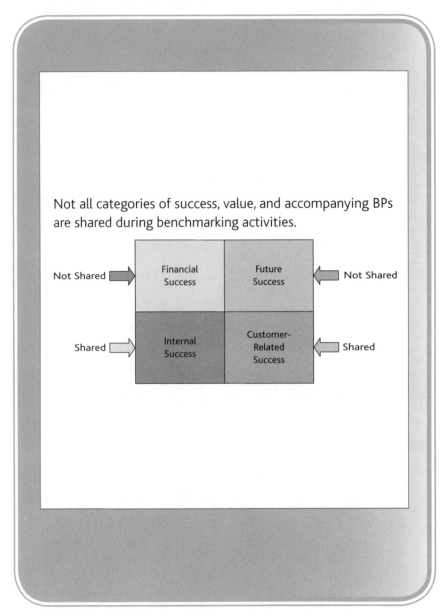

Not all categories of success, value, and accompanying BPs are shared during benchmarking activities.

As seen in the preceding illustration, not all categories of success or value are shared during benchmarking. There are several reasons for this:

- The information is regarded as proprietary knowledge.

- You must provide the sharing organization with the same quality information that they provide you. Equal sharing is essential.

- The organization performing the benchmarking views you as a competitor.

- There is uncertainty and concern about what you will do with the information.

- The organization has had a bad experience previously with benchmarking and does not want to repeat mistakes.

Companies may be willing to share information about internal success and customer-related success because it is less threatening than releasing information about financial success and future success. Releasing information about future success could result in disclosure of the company's strategic plans.

INTELLECTUAL PROPERTY COST VERSUS VALUE

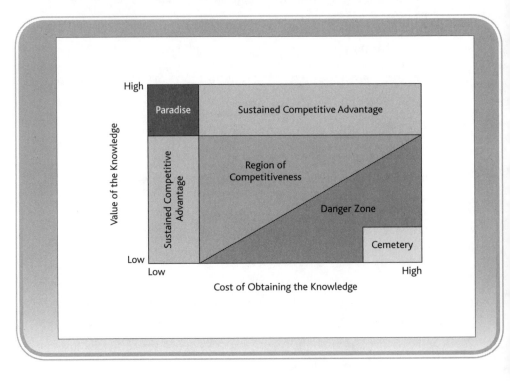

Developing or obtaining intellectual property, either internally or through benchmarking, and whether it is considered to be a BP or value metric, requires management commitment and funding. If the value of the information is considered to be high and the cost of obtaining the information is low, this is a very desirable situation and the return on investment (ROI) can be significant. One BP can save a company millions of dollars in the long term.

As the cost of obtaining the information increases and the value of the information decreases, we enter a "Danger Zone." In the "Cemetery" box, failure may be imminent because we would be spending a significant amount of money to capture information that is of little value or no value at all. Doing this for a prolonged time could be disastrous.

IMPLEMENTATION FAILURES

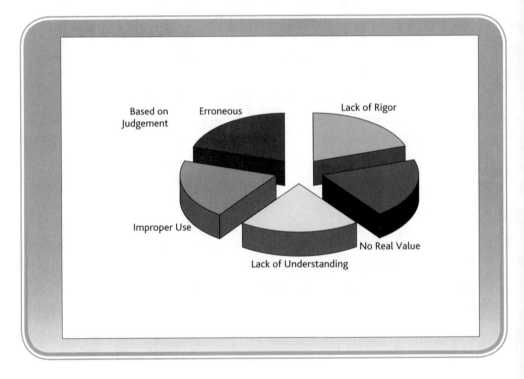

Sometimes, what appears to be very positive on paper cannot be implemented as described or made to work within your company. Some BPs are based on the corporate culture and how the employees use the BPs. Your company may have a completely different culture, which will affect how BPs are accepted and used. A few examples might be:

- The other company's BPs were based on erroneous judgment or a one-time situation, and you didn't recognize it until implementation.

- The BP lacks vigor and is too simplistic.

- The implementation of the BP does not provide any real value to your company.

- Your understanding of the BP was faulty or incomplete.

- You see no value in the BP because implementation was completed improperly.

- The internal operating values of the company established by management conflict with the BP.

Chapter

THE VALUE
CONTINUUM

THE TIMING OF VALUES

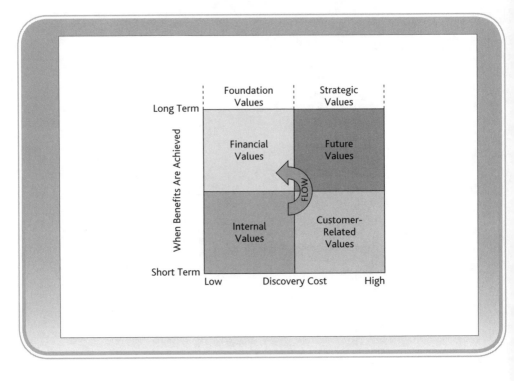

A question that is frequently asked about values is, "How does a company determine the timing associated with the defined values?" The order for a typical company is shown in the illustration on the previous page. Internal values come first because the company must have some type of repetitive internal success just to stay in business. Once internal values are achieved, customer-related values follow. A company may be willing to incur financial losses for a short period of time to develop a meaningful customer base.

In third place are future values. The future value may be based on the need to satisfy customers in the long term, and, once again, the company may be willing to absorb near-term losses to achieve greater value later. Finally, we come to financial values. A company cannot absorb losses indefinitely and remain in business. The company must eventually become successful financially in order to remain solvent in the long term.

THE VALUE CONTINUUM

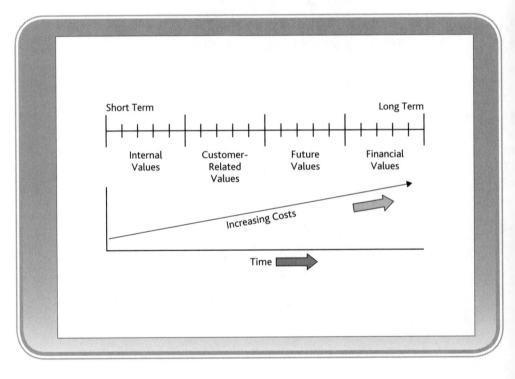

The preceding illustration is a slightly different representation of the timing of the values. As expected, the cost can increase significantly as we move across the continuum. And, as stated in the previous illustration, the company must eventually obtain a revenue stream to cover the costs.

BARRIERS ALONG THE CONTINUUM

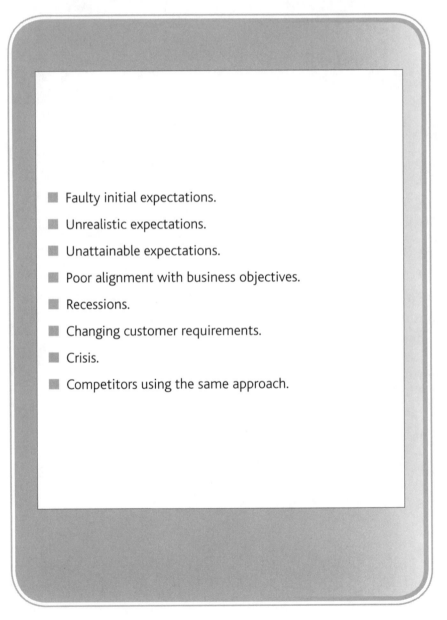

- Faulty initial expectations.
- Unrealistic expectations.
- Unattainable expectations.
- Poor alignment with business objectives.
- Recessions.
- Changing customer requirements.
- Crisis.
- Competitors using the same approach.

The road to success is not always straight and narrow. There are barriers along the continuum, and these barriers can elongate the time needed to achieve the value associated with the cost. Some barriers can be compensated for as the company matures, but trying to overcome several barriers occurring at the same time may be impossible. Understanding and preparing for these potential barriers will minimize their impact and, in some cases, accelerate the realization of value.

"Chance favors the prepared mind."

LOUIS PASTEUR - FRENCH CHEMIST AND MICROBIOLOGIST

ACTIVITIES TO SPEED UP THE VALUE CONTINUUM

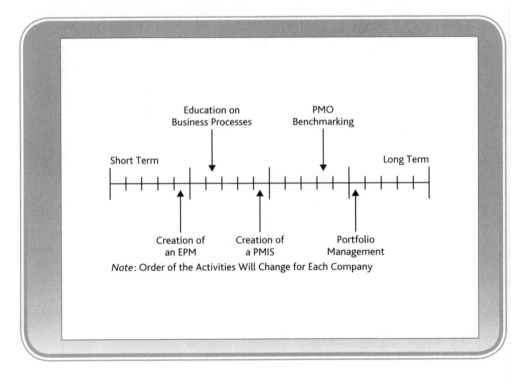

There are some activities that can be undertaken to speed up the value continuum. The order of the added activities will change from company to company, as well as the timing of implementation. For example, the early establishment of a project management office (PMO) during the project management maturity process can accelerate the realization of value. The creation of an enterprise project management (EPM) methodology can help in obtaining all four forms of value, and therefore should be accomplished as quickly as possible.

THE VALUE CONTINUUM AND THE PROJECT MANAGEMENT MATURITY MODEL

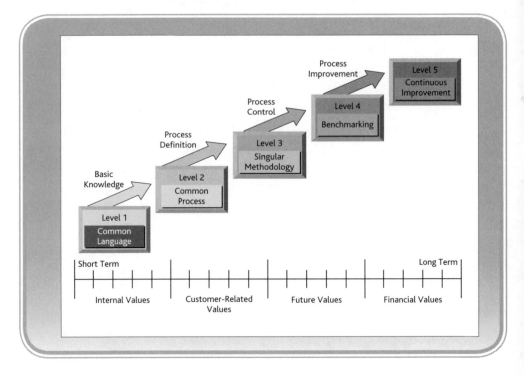

The value continuum can also be used in conjunction with maturity models for project management. The preceding illustration maps the continuum against the Project Management Maturity Model (PMMM).* The PMMM is one of several project management maturity models in the marketplace. The PMMM has five levels:

- **Level 1:** Providing training programs on project management so that people become familiar with project management terminology, and possibly have some people become Project Management Professionals (PMPs).

- **Level 2:** The people you have trained begin the development of processes for project management.

- **Level 3:** The processes developed in Level 2 can be put together into an EPM system.

- **Level 4:** Perform benchmarking on project management.

- **Level 5:** Decide what information, if any, from the benchmarking studies should be implemented into the EPM system as part of continuous improvement.

---

*For additional information of PMMM, see Harold Kerzner, *Strategic Planning for Project Management Using a Project Management Maturity Model*, 2nd ed. Hoboken, NJ: John Wiley & Sons, © 2006.

VALUE MANAGEMENT LIFE-CYCLE PHASES

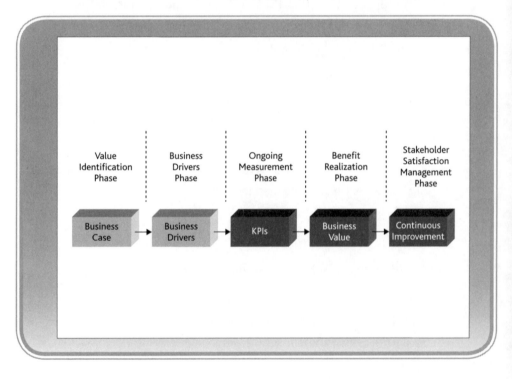

It is possible that the traditional EPM systems with traditional life-cycle phases may need to be changed if achieving value becomes the ultimate goal rather than simply meeting the specifications of the triple constraint. The preceding five life-cycle phases may become the overall structure for projects, and the traditional life-cycle phases and accompanying methodology may be included within each of these phases. In other words, we may end up with a methodology within a methodology.

Traditional EPM systems focus on a well-defined set of objectives, which most likely will not change over the life of the project, and where the target is to complete the project within the specified project triple constraint. But when value becomes the target, we may need to allow the project objectives to change as the project progresses through each phase of the project life cycle because of the necessity to remain flexible to the needs of the customer and to the changes in the business environment. True value will ultimately be achieved through controlled adaptation. The life-cycle phases shown will be discussed in the next several illustrations.

VALUE IDENTIFICATION PHASE: BUSINESS CASE

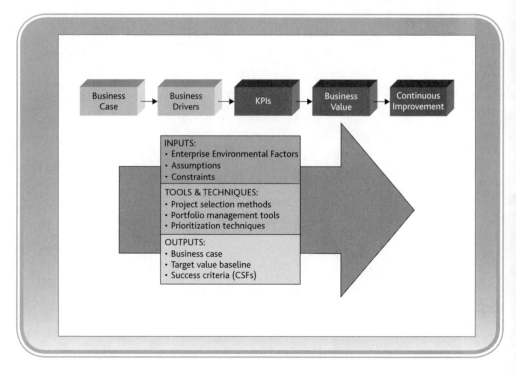

The first life-cycle phase includes the identification of the factors that would be used in the development of the business case that will focus on obtaining the value. The appropriate selection methods would be considered along with other criteria that would provide a confidence level that value will be achieved. The output would be the business case and the target value baseline against which progress will be measured. In this phase, the development of the *target value baseline* and the business case could be considered a project in itself using the traditional EPM methodology.

BUSINESS DRIVERS PHASE: BUSINESS DRIVERS

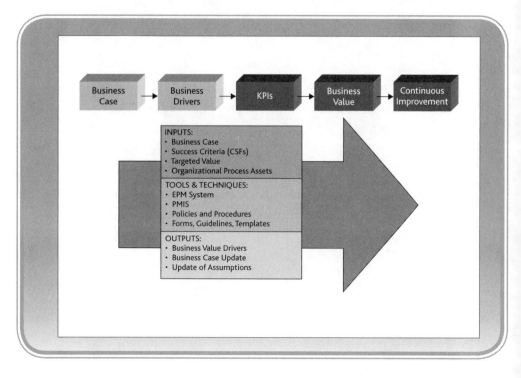

Business Case → Business Drivers → KPIs → Business Value → Continuous Improvement

INPUTS:
- Business Case
- Success Criteria (CSFs)
- Targeted Value
- Organizational Process Assets

TOOLS & TECHNIQUES:
- EPM System
- PMIS
- Policies and Procedures
- Forms, Guidelines, Templates

OUTPUTS:
- Business Value Drivers
- Business Case Update
- Update of Assumptions

The second life-cycle phase is the identification of the business value drivers. The value drivers, also called performance drivers, identify how you will deliver the business case. The value drivers can include the EPM system, project management information system (PMIS), and other organizational process assets that will be used.

The business value drivers are the tools, techniques, and processes that will be employed to help create the ultimate targeted value. Business value drivers could also be identified in terms of resources. As an example, an employee with critical skills must be assigned to the project because that resource or required skill is a business value driver. Business value drivers can change from project to project.

MEASUREMENT PHASE: KEY PERFORMANCE INDICATORS

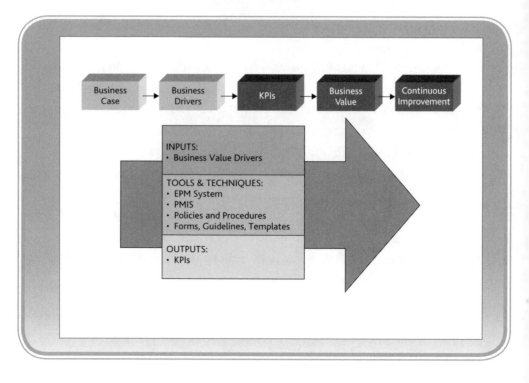

The third life-cycle phase is the measurement phase. In this phase, we identify the key performance indicators (KPIs) that should be measured in connection with the business value drivers. The KPIs can differ from project to project. The KPIs could be the quality of the resources assigned, the interim deliverables, customer satisfaction rating reports, variances against baselines, usage of specific tools, and usage of project-related best practices (BPs).

A KPI can have a target, which is an exact value that the KPI should achieve. It can also have ranges against which to track the KPI. Ranges can be either a percentage of the target value or an actual value.

The difficulty is to determine:

- How many KPIs should be considered?

- Which ones are most important?

- How to measure and evaluate the KPIs?

- How frequently to perform the measurement?

VALUE REALIZATION PHASE: VALUE (BENEFITS)

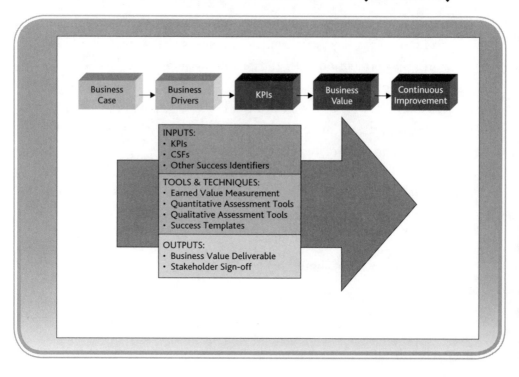

The fourth life-cycle phase is the obtaining of the business value and the eventual sign-off by the stakeholders. This is the accomplishment of the deliverables. This life-cycle phase can be quite long, depending on the type of project and the deliverables.

If the value cannot be verified at the end of the phase, then an investigation must be made as to whether or not the assumptions had changed or if there were any other significant changes that may have affected the project. If the value is verified as achieved, then the appropriate stakeholder(s) should sign off and accept the deliverables.

CUSTOMER SATISFACTION MANAGEMENT PHASE:
CONTINUOUS IMPROVEMENT

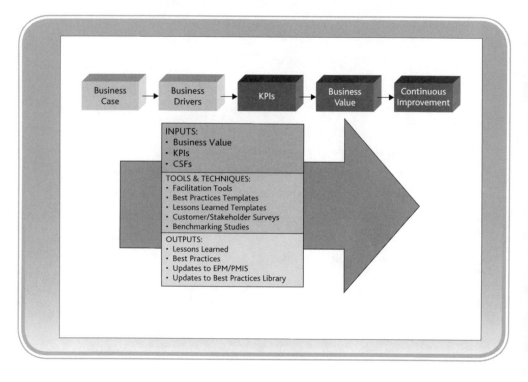

The last life-cycle phase is customer satisfaction management. This is the phase where BPs and lessons learned are captured. The participants in this phase could include the project manager, sales and marketing personnel, both internal and external customers, facilitation personnel with expertise in extracting BPs, and project office personnel.

If the project is a long-term effort, then the capturing of BPs and lessons learned may be accomplished intermittently throughout the project rather than just at the project's completion. There may be a template as part of the EPM system to extract BPs. Also, since BPs and lessons learned can be found in both successes and failures, this phase may be performed for projects that are terminated even though no value has been developed.

Some projects must be terminated prior to the completion of the full life cycle. There are performance reports that can provide us with some indication that a project should be terminated. This is shown in the next chapter.

ASSIGNING VALUE THROUGH OBJECTIVES

TYPES OF PERFORMANCE REPORTS

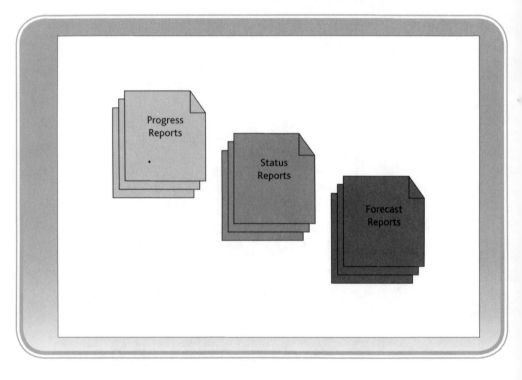

Performance reporting is essential for effective decision making to take place. In general, there are three types of performance reports:

- **Progress reports:** These reports describe the work accomplished to date. This includes:

    - The planned amount of work up to the timeline of the report

    - The actual amount of work accomplished up to the timeline

    - The actual cost accumulated up to the timeline

- **Status reports:** These reports indicate the status by comparing the progress to the baselines and determining the variances. This includes:

    - The schedule variance up to the timeline of the report

    - The cost variance up to the timeline

- **Forecast reports:** The progress reports and status reports are snapshots of where we are today. The forecast reports indicate where we will end up. This includes:

    - The expected cost at the completion of the project

    - The expected time duration or date at completion of the project

There are other items that can be included in these reports. However, our main concern, as seen in the next illustration, is with the forecast reports.

BENEFITS AND VALUE AT COMPLETION

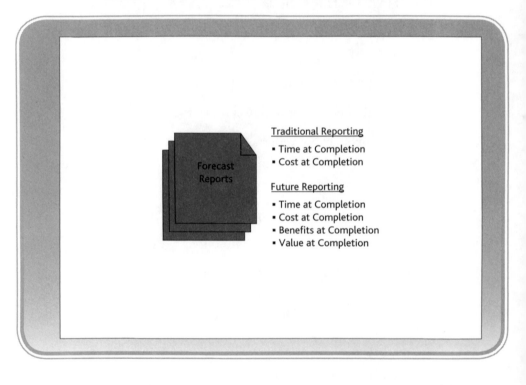

Traditional forecast reports provide information about the time and cost expected at the completion of the project. This data can be calculated from extrapolation of trends or formulas. Unfortunately, this data may not be sufficient to provide management with the necessary information to make effective business decisions and to decide whether to continue on with the project or consider termination.

Two additional pieces of information are necessary: the expected benefits at completion and the expected value at completion. Most earned value measurements systems in use today do not report these two additional pieces of information, probably because there are no standard formulas for them.

The benefits and value at completion must be calculated periodically. However, based on which life-cycle phase you are in, there may be insufficient data to perform the calculation quantitatively. In such cases, a qualitative assessment of benefits and value at completion may be necessary, assuming, of course, that information exists to support the assessment. Expected benefits and value are more appropriate for business decision making and usually provide a strong basis for continuation or cancellation of the project.

DETERMINING BENEFITS (VALUE) AT COMPLETION

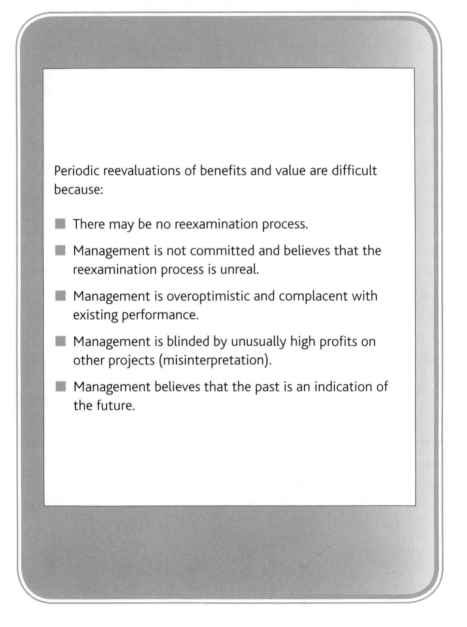

Periodic reevaluations of benefits and value are difficult because:

- There may be no reexamination process.

- Management is not committed and believes that the reexamination process is unreal.

- Management is overoptimistic and complacent with existing performance.

- Management is blinded by unusually high profits on other projects (misinterpretation).

- Management believes that the past is an indication of the future.

Some form of reevaluation or reexamination process must exist to determine if an organization is working on the right projects and if the benefits are still achievable. But, as can be seen from the preceding illustration, there may be roadblocks that discourage a reexamination process. As an example, an executive may have funded a "pet" project and may be afraid of the realities that would be discovered during the reexamination process.

Reexamination processes need not be accomplished at the same time as the end-of-phase review meetings that are part of an enterprise project management (EPM) methodology. They may be accomplished monthly, based on availability of information, at the discretion of the exit champion, assuming an exit champion exists, or when a significant change occurs in the business or the economic environment.

Reevaluation processes can be accomplished based on qualitative rather than quantitative information. Sometimes, quantitative financial numbers are not available and decisions must be made on qualitative factors.

ESTABLISHING THE BUSINESS OBJECTIVES

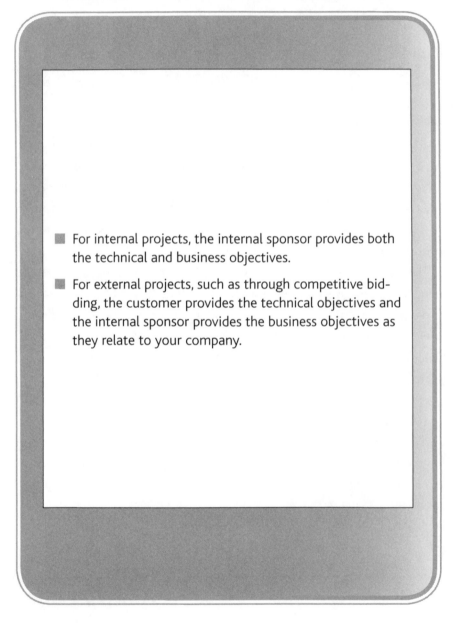

- For internal projects, the internal sponsor provides both the technical and business objectives.

- For external projects, such as through competitive bidding, the customer provides the technical objectives and the internal sponsor provides the business objectives as they relate to your company.

How do we use the SMART rule for business objectives such as:

■ Increase market share.

■ Become a market leader.

■ Lower our costs.

■ Develop a reputation for superior service.

■ Have wider profit margins.

■ Become recognized as a blue-chip company.

■ Develop better credit ratings.

■ Have faster earnings/revenue growth.

Business objectives can be at a higher level and considerably vaguer than standard project objectives. We generally apply the SMART rule to project objectives, namely:

- **S**pecific
- **M**easurable
- **A**ttainable
- **R**ealistic or **R**elevant
- **T**ime-bound

The application of the SMART rule works reasonably well when the objectives are well defined and well understood. Reexamination of the value and benefits of the project is made by a comparison against the objectives. But business objectives can be significantly vague, and the SMART rule may be more difficult to apply. It may be necessary to develop another rule similar to the SMART rule for the project's business objectives—possibly SOUND:

- **S**trategic
- **O**ptimistic
- **U**seful
- **N**ecessary
- **D**eliverable

ESTIMATING APPROACHES

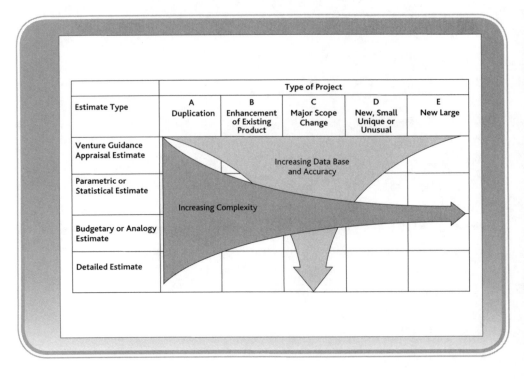

Estimate Type	Type of Project				
	A Duplication	B Enhancement of Existing Product	C Major Scope Change	D New, Small Unique or Unusual	E New Large
Venture Guidance Appraisal Estimate					
Parametric or Statistical Estimate					
Budgetary or Analogy Estimate					
Detailed Estimate					

Increasing Data Base and Accuracy

Increasing Complexity

The achievement of the objectives, and the ultimate value, is heavily dependent on the quality and reliability of the estimates being used. Unfortunately, organizations are now working on projects that have a greater degree of uncertainty and risk and, as expected, our ability to estimate these projects is challenged, and in many cases effective estimating techniques are not utilized.

The preceding illustration illustrates various estimating approaches. Many of the projects currently in progress across most industries are categorized as Type D and E projects. These projects are most often estimated using the venture guidance appraisal approach, which has the lowest degree of accuracy. This is one of the primary reasons that we are unsure as to what the final value will be.

PROJECT PLANS

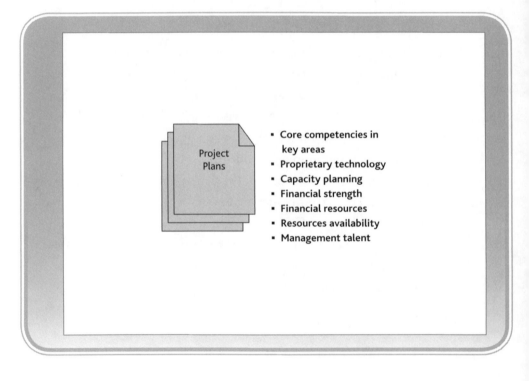

Project managers are expected to develop a project plan through the input and expertise of the project team, and the plan is generally targeted for the completion of the project's objectives. But this assumes that we have the traditional types of project objectives defined.

The preceding illustration shows several of the items that we look at for the development of a project plan. All of these items generally look at the availability and quality of corporate resources that will be used. Project plans are reasonably easy to develop and update should economic conditions change. Business plans, however, can be difficult to update.

BUSINESS PLANS

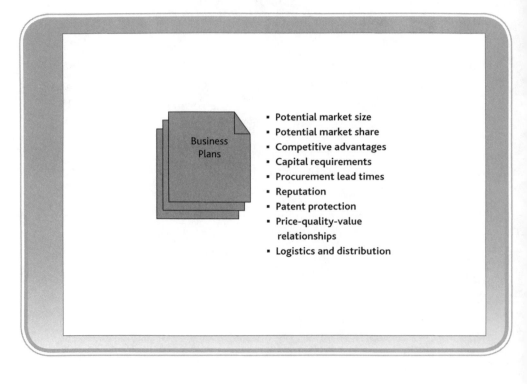

When business value is to be included as part of the success criteria for a project, the project manager may find it necessary to develop a business plan for the project as well as a project plan. Business plans are based on the items in the preceding illustration and may not be as easy to develop as the traditional project plan. There are several reasons why the development of business plans may be difficult:

- Project plans are more concrete, whereas business plans deal with the abstract

- Project plans deal with a much shorter time frame than business plans

- The baselines in project plans are easier to develop and measure than the baselines in business plans

- Baselines in a project plan will undergo less frequent changes than baselines in a business plan

There are other differences, but we can see the necessity for possibly developing both a traditional project plan and a business plan. Rolling wave planning may be the best way to deal with business plans.

CANCELING PROJECTS

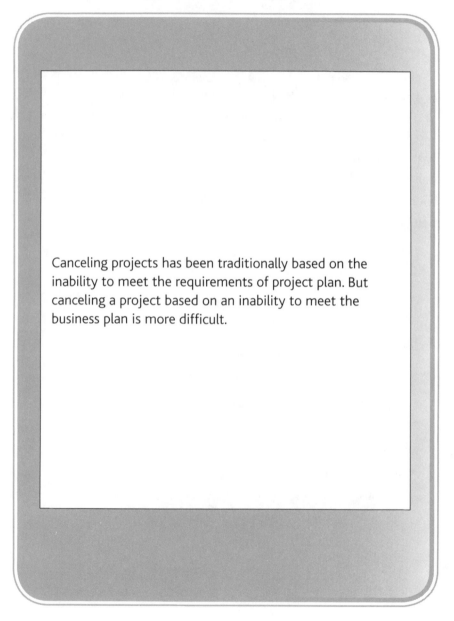

Canceling projects has been traditionally based on the inability to meet the requirements of project plan. But canceling a project based on an inability to meet the business plan is more difficult.

If they haven't done so already, companies can be expected in the future to establish some sort of criteria for the cancellation of projects. This may be accomplished with an executive steering committee or by creating the position of "exit champion." Historically, very few projects were canceled and many projects were ultimately completed as planned but provided no real value to the company. The reason for this was that the business plan had changed and there was no alignment with the project plan. The project plan was left intact and executed without regard to the changes in the business plan.

The cancellation criteria should be based on both the business plan and the project plan. Because business plans and business objectives are focused on a longer time frame and are based on higher-level, less reliable estimates, there is a tendency to postpone making a decision to cancel a project based solely upon changes to the business plan. This can be a costly mistake.

MARRYING PROJECT AND PROGRAM MANAGEMENT

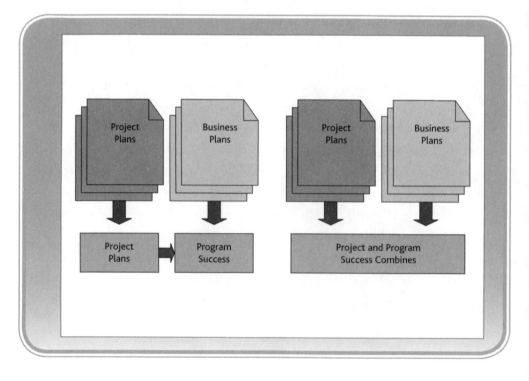

In the very near future, if not already in place in some organizations, we can expect the definition of true success to include project success criteria and business success criteria, and possibly be combined into one definition of success. But this does not mean that project and program management will be combined into a single discipline. It simply means that companies will add a business component into the definition of project success.

VALUE LEADERSHIP AND SENIOR MANAGEMENT

THE EVOLUTION OF LEADERSHIP

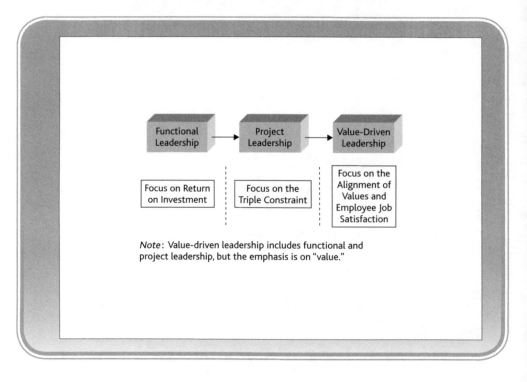

The preceding illustration shows the evolution of value-driven leadership. The focus is more on the alignment of values than the triple constraint. Value-driven leadership will change changed our definition of *success*.

The expected outcome of value-driven leadership includes:

- An alignment of values with both business and project objectives

- Greater profitability for the performing organization and greater benefit realization for the receiving organization

- Outperforming the competitors

- It will be easier to implement organizational changes

- Lower employee turnover, especially workers with critical skills

- Reduction in operating costs

- Higher quality of each project deliverable and at every level and business unit in an enterprise

- Willingness to cancel nonperforming of non-value-added projects

MEASUREMENTS AND TRIGGERS

What are the measurements or triggers indicating that value-driven leadership, including project management, is working as expected?

Fortunately, there are measurements that show that value-driven leadership is working. There are hard or quantitative measurement and soft or qualitative measurements. The hard measurements include:

- Return on investment (ROI) results

- Net present value results

- Internal rate of return (IRR) results

- Reduction in costs

- Higher levels of measurable quality

The softer measurements include:

- Improved teamwork

- Fewer conflicts requiring executive attention

- Better decision making

- More accurate information

- Higher morale

- Higher levels of customer satisfaction

- Better brand recognition and image

WHAT EXECUTIVES WANT TO HEAR

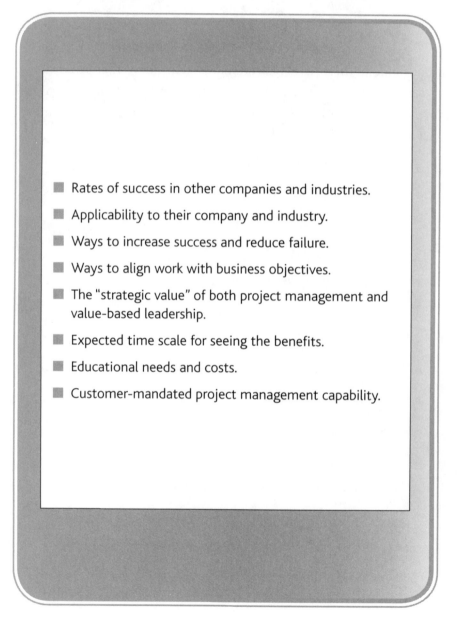

- Rates of success in other companies and industries.
- Applicability to their company and industry.
- Ways to increase success and reduce failure.
- Ways to align work with business objectives.
- The "strategic value" of both project management and value-based leadership.
- Expected time scale for seeing the benefits.
- Educational needs and costs.
- Customer-mandated project management capability.

Selling value-driven leadership to executives has many of the same characteristics as selling project management to executives. Value-driven leadership includes both functional (or traditional) leadership and project management leadership. There are both project management components and business components.

We must remember that it has taken us years to get executives to believe that project managers should be empowered to make project-related decisions. Now we are asking executives to believe that project managers can make business-related decisions and that project managers understand the business and the consequences of their decisions.

Some executives may view project management as a severe threat to their authority and decision-making responsibility with regard to profit and loss. Giving a project manager business-related decision making may be seen as a threat to the size of the executive's year-end bonus, especially if the bonus must be shared with the project manager.

In the life-cycle model shown previously, it was explained that the first life-cycle phase required an identification of the business case. This meant that both the technical and business objectives had to be defined.

Project managers are generally more accustomed to working with technical objectives rather than business objectives. If business objectives have been established, then the information most likely was provided by the internal sponsor. For projects that are internal to the organization, the business objectives are readily understood. For projects external to your organization, the supplier organization's business objective and the customer's business objective could be misaligned or can be in disagreement. As an example, the customer's reason for awarding an organization a project may be to develop a product or service that they could market and sell to generate revenue. Your company's business objective might be simply to keep your people employed or to take advantage of underutilized capacity in your company.

If project managers are expected to make business decisions in addition to project decisions, then the project managers must be fully aware of and understand the business objectives. Unfortunately, what many organizations refer to as business objectives look more like business goals than business objectives and may not be defined quantitatively.

CRITICAL ISSUES FOR THE SELLING PROCESS

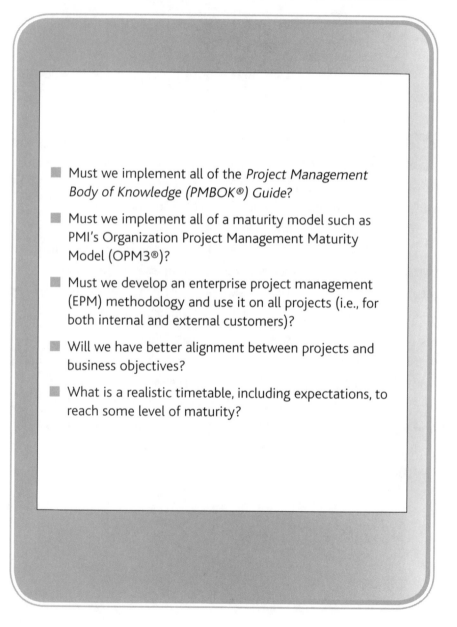

- Must we implement all of the *Project Management Body of Knowledge (PMBOK®) Guide*?

- Must we implement all of a maturity model such as PMI's Organization Project Management Maturity Model (OPM3®)?

- Must we develop an enterprise project management (EPM) methodology and use it on all projects (i.e., for both internal and external customers)?

- Will we have better alignment between projects and business objectives?

- What is a realistic timetable, including expectations, to reach some level of maturity?

The preceding example shows some of the critical issues that need to be considered. Executives are often fearful of having to implement processes that may not be needed, especially if the *PMBOK® Guide* is used as the basis for a project methodology. The *PMBOK® Guide* is, just as the name implies, a guide, not a set of policies and procedures that must be followed exactly.

Good project management methodologies allow for the inclusion of earned value measurement tools to support the budgeting and tracking of project costs. This is often referred to as horizontal accounting. Some executives do not want to see horizontal accounting implemented and may not want to hear that their original estimates were wrong or way off the mark.

Executives are also concerned with the time needed to reach maturity. Executives are fearful of committing vast resources to a maturity process that may have no end date in sight. The focus here should be on continuous gradual improvement with value as the key driving factor.

THREATS THAT EXECUTIVES FACE

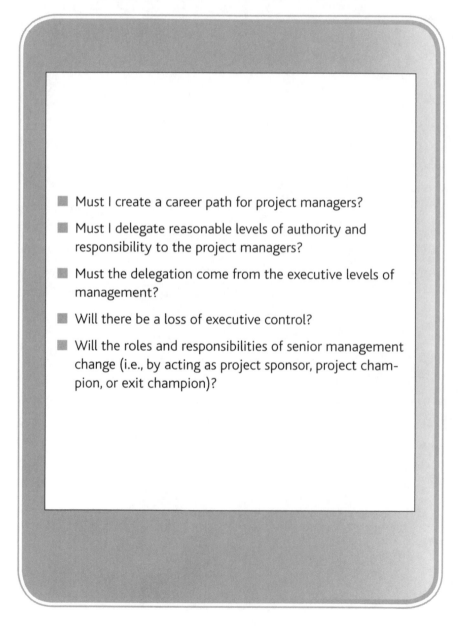

- Must I create a career path for project managers?
- Must I delegate reasonable levels of authority and responsibility to the project managers?
- Must the delegation come from the executive levels of management?
- Will there be a loss of executive control?
- Will the roles and responsibilities of senior management change (i.e., by acting as project sponsor, project champion, or exit champion)?

There are also organizational threats that executives face. Defining project management a career path position may seem to be the appropriate thing to do, but there are a few items to consider:

- There may be a business risk if project managers will be given the responsibility for profit and loss
- The project managers may become more influential than some of the executives

Another fear is loss of control. Mature project management advocates the decentralization of authority and decision making. For some executives, this will most certainly be viewed as a threat. This problem can be effectively resolved by clearly defining the role of the executive or project sponsor and the roles of the project manager along a set of clearly defined expectations.

PROJECT MANAGEMENT SUCCESS VERSUS MATURITY

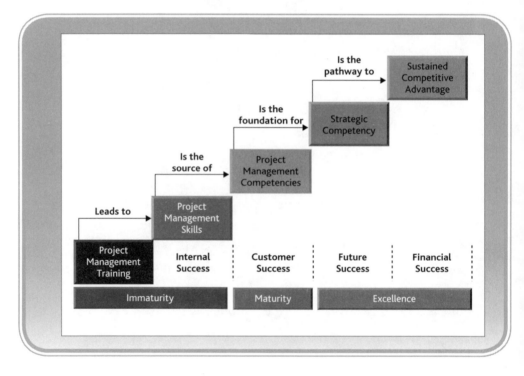

Capturing value and having some successes is no guarantee that the organization has achieved excellence in project management. Maturity can be defined as a business culture in which repeatable processes are in place and the appropriate support systems have been developed that generate a continuous stream of deliverables. Excellence is achieved when these deliverables provide sufficient value from the stakeholder perspective that the future of the company is assured.

When companies recognize the need for excellence in project management, they position project management as a career path and view it as a strategic competency. The company then capitalizes on its strategic competencies through effective advertising and promotion of project management capabilities to provide the company with a sustained competitive advantage.

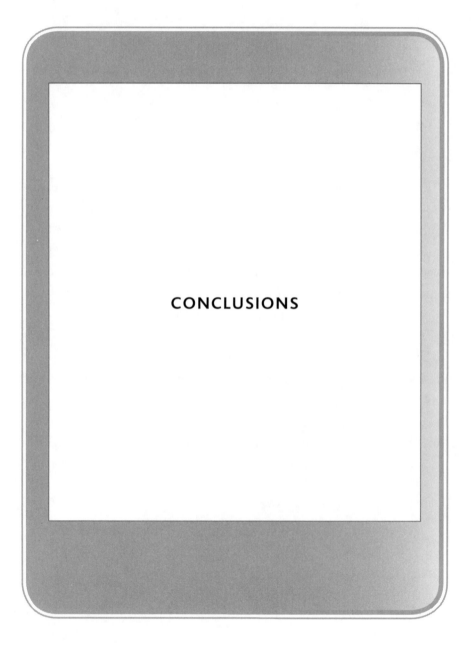

CONCLUSIONS

The following list summarizes the conclusions provided in this book:

- Project managers will become more involved with making business decisions related to projects.

- Project success will be defined in terms of internal factors such as process improvement, financial factors such as ROI or cost reduction, success, customer-driven success such as customer satisfaction and the achievement of expected value, and future success such as add-on business and greater market share.

- Each of the four components of success related to a project will contain a business component defined in terms of value.

- The concept of managing for the value expected in the project rather than the triple constraint will become the major focus of an enterprise.

- Because we are working on projects that have much greater risk in terms of threats and opportunities than ever before, the final achieved value of the project may be significantly different than what was expected by the stakeholders.

- Project managers will be transformed into business managers with an equal emphasis on project success and business success.

INDEX

A

Antares Solutions, 156–157
Asea Brown Boveri (ABB), 160–161
Automotive suppliers sector,
    representation, 170–171

B

Banking sector, representation, 172–173
Benchmarking, 203
  values, sharing, 204–205
Best practices (BPs), 182–183
  acceptance, 209
  capture, 14
  disclosure/discovery, 192–193, 196–197
  usage, 229
Budget, control (postulate), 34
Bureaucracy, project management
    derivative, 91
Business
  decision making, quality (absence), 53
  drivers phase, 226–228
  plans, 250–251
  results/expectations, 55
Business objectives
  establishment, 242–243
  SMART rule, usage, 244–245
  usage, 243
Business-related changes,
    implementation (absence), 46
Business value
  achievement, postulate, 44
  customer perception, postulate, 42–43
  obtaining, 231
  types, 68–69

C

Cancellation criteria, 253
Capability Maturity Model (CMM), 93
Capital improvements project, example,
    158–159
Capital market stakeholders, 107
Commitment, action, 99
Commodity products (manufacturing)
    sector, representation, 174–175
Companies, representation, 176–179
Competitive bidding activities, contractor
    emphasis, 11
Completion benefits/value, 238–239
  determination, 240–241
Compliance, action/tendency, 99
Computer Associates (CA) Technology
    Services, 164–165
Conformity, predictability
    (relationship), 83
Continuous improvement, 101, 221,
    232–233
Convergent Computing (CCO),
    166–167
Cost overrun, acceptance, 119
Critical success factors (CSFs), 183–189
  metrics, establishment, 187
Customer-related success, 131
Customer-related values, 146–147, 213
  failure, reasons, 154–155
Customers
  appeasement, 53
  requirements, achievement, 11
  satisfaction management phase,
    232–233

D

Debriefing pyramid, 194–195
Decision trees, 145
Deliverable
  production, request/planning, 33
  value, absence, 47
Denver International Airport (DIA)
  cost, problems, 116–119
  stakeholders, 118

E

Earned value measurement techniques,
    focus, 47
Education, importance, 93
End-of-phase review meetings, 241
Engagement project management,
    emphasis, 77
Engineers, project manager
    designation, 21
Enterprise project management (EPM)
    methodologies
  adoption, 13
  business processes, 12
  creation, 133, 219
  development, 7, 141, 245
  focus, 6
  inclusion, 38
  understanding, 39
  usage, advantages, 83
Estimates, usage, 247
Executives
  concerns, 262–263, 265
  organizational threats, 267
  project sponsor training, 9
  threats, 266–267
External projects, 242

F

Financial success, 131
Financial values, 137, 142
  failure, reasons, 150–151
  focus, 143
Forecast reports, 237–239
Foundation values, 68–69, 137
Functional (traditional) leadership, 263
Future success, 131
Future values, 144–145, 213
  failure, reasons, 152–153

G

General Electric (Plastics Group), 158–159

H

Hard measurements, 261

I

Implementation failures, 208–209
Information technology, example, 157, 165
Innovation values, 68–69, 137
Intellectual property
  cost, value (contrast), 206–207
  impact, 115
Internal collaboration, focus, 85
Internal projects, 242
Internal rate of return (IRR), 55, 145
  results, 261
Internal success, 131
Internal values, 137, 140, 213
  failure, reasons, 148–149
International Organization for
    Standardization (ISO 9000), 93
  certification, 175

J

Job descriptions, company description, 73

K

Key performance indicators (KPIs),
    183–189, 228–229
  identification, 229
  metrics, establishment, 187

L

Leadership, evolution, 258–259
Life-cycle phase, 222–233
  business case, identification
    (requirement), 243
Lisa Computer (Apple), 112
  project stakeholders, 114
  value, 113, 115
Long-term financial failure, impact, 151
Long-term financial health, 143
Long-term partnership agreements,
    expense, 23

M

Management, conflicts, 123
Measurement phase, 228–229
Metrics, change (requirement), 188–189

Middle management, technical knowledge (possession), 31
Mission critical issues, 59
Motorola, 167–168
Multidirectional leadership, effectiveness, 95

O

Organizational Project Management Maturity Model (OPM3), 93, 245
Organizational stakeholders, 107

P

Performance
  reports, types, 236–237
  Six Sigma levels, 175
Planning, focus, 79
Price, postulate, 40–41
Proactive management, impact, 87
Process improvement, 169, 177
Product enhancements, 177
Product/market stakeholders, 107
Program management, project management (combination), 254–255
Program success, definition, 54–55
Progress reports, 237
Project center of excellence, function, 101
Project management
  benchmarking, 202–203
  business process, 10
  discipline, 9
  evolution, 50
  failure, 2
  formality/informality, 89
  growth, importance, 163
  importance, 29
  knowledge, 200–201
  leadership, 263
  methodologies, 91
    components, 13
  perspective, 8
  practices, maturity, 38
  program management, combination, 254–255
  success, maturity (contrast), 268–269
  usage, limitations, 29
  value
    conflicts, 124–125
    types, application, 65
  viewpoints, change, 16–30

Project Management Body of Knowledge (PMBOK), 264–265
Project management information system (PMIS), 188–189, 227
Project Management Maturity Model (PMMM), 221
  value continuum, relationship, 220–221
Project management office (PMO), 143
  establishment, 219
  function, 101
  importance, 193
  involvement, 190–191
  usage, 31
Project Management Professionals (PMPs), 221
Project managers
  authority/power, requirement, 75
  business knowledge, limitation, 25, 97
  business managers, transformation, 271
  core competency models, 73
  execution specialists, 19
  input, request, 17
  involvement, 5, 271
  mistrust, 71
  negotiation, 75
  organizational resources, 17
Project-related best practices, usage, 229
Projects
  cancellation, 252–253
  completion, 35
    constraint, 36
    triple constraint, impact, 62
  definition/understanding, problems, 2–3
  execution, postulate, 32
  plans, 46, 248–249, 251
  risk, 271
    project management avoidance, 79
  success, defining, 131
  value perceptions, 126–127
Project success
  definition, 54–55, 271
    modification, 25
  program/business success, differentiation, 27

Q

Quadrant, decision making, 138–139
Quality, importance, 41

R

Reevaluation processes, 241
Reexamination processes, 241
Request for proposal (RFP), issuance, 15
Return on investment (ROI), 25, 55,
    66–67, 93
  results, 261
  significance, 207

S

Security, achievement, 81
Sensitivity analysis, 145
Soft measurements, 261
Specific Measurable Attainable Realistic/
    Relevant Time-bound (SMART)
    rule, application, 244–245
Stakeholders
  classification, 106–107
  needs, balancing, 120–121
  perceptions, 104–105
Statement of work (SOW), 4–5
Status reports, 237
Strategic objectives, value alignment, 37
Strategic Optimistic Useful Necessary
    Deliverable (SOUND) rule,
    usage, 245
Strategic project management,
    importance, 97
Strategic values, 68–69, 137
Success
  achievement, 44, 62
  categories, 132
  components, 130, 135, 184, 271
  criterion, triple constraint, 52
    redefining, 56
  definition, 58–59, 187, 259
    change, 50–51
    value, inclusion, 63
  focus, 39
  project manager definition, 77
Sydney Opera House (Australia)
  cost overruns/scope changes, 111
  project stakeholders, 110
  value, perception, 108–109

T

Target value baseline, development, 225
Technical objectives, usage, 243
Time/cost/performance, triple constraint, 57

Timeliness, importance (postulate), 34
Total business solutions, emphasis, 77
Total product offering, 41
Triple constraint, 52–57, 135
  focus, 133
  success criteria, redefining, 56
Trust, 71, 75

V

Value
  achievement, 43, 135
  addition, proactive management
    (impact), 87
  categories, 134–135
  change, 70–72
  conflicts, 122–123
  continuum, 214–215
    acceleration, activities, 218–219
    barriers, 216–217
    project management maturity model,
      relationship, 220–221
  definition, 45
    ROI, impact, 67
  drivers, example, 160–161
  expectations, establishment
    (failure), 47
  focus, 37
  identification phase, 224–225
  management, 271
    life-cycle phases, 222–223
  metrics, redefining, 186–187
  obtaining, success levels, 198–199
  perception, 125
  project management, 119
  realization phase, 230–231
  receipt, postulate, 40–41
  timing, 212–213
  types, 64–65
Value-based project management,
    goal, 183
Value-driven leadership
  expected outcome, 259
  measurements/triggers, 260–261
  selling process, 263
    critical issues, 264–265
Value-oriented benchmarking, 203

W

Westfield Group, 162–163